T0324910

BASIC CATEGORY THEORY

At the heart of this short introduction to category theory is the idea of a universal property, important throughout mathematics. After an introductory chapter giving the basic definitions, separate chapters explain three ways of expressing universal properties: via adjoint functors, representable functors and limits. A final chapter ties all three together.

The book is suitable for use in courses or for independent study. Assuming relatively little mathematical background, it is ideal for beginning graduate students or advanced undergraduates learning category theory for the first time. For each new categorical concept, a generous supply of examples is provided, taken from different parts of mathematics. At points where the leap in abstraction is particularly great (such as the Yoneda lemma), the reader will find careful and extensive explanations. Copious exercises are included.

Tom Leinster has held postdoctoral positions at Cambridge and the Institut des Hautes Études Scientifiques (France), and held an EPSRC Advanced Research Fellowship at the University of Glasgow. He is currently a Chancellor's Fellow at the University of Edinburgh. He is also the author of *Higher Operads, Higher Categories* (Cambridge University Press, 2004), and one of the hosts of the research blog, The *n*-Category Café.

CAMBRIDGE STUDIES IN ADVANCED MATHEMATICS

Editorial Board:
B. Bollobás, W. Fulton, A. Katok, F. Kirwan, P. Sarnak, B. Simon, B. Totaro

All the titles listed below can be obtained from good booksellers or from Cambridge University Press. For a complete series listing visit: www.cambridge.org/mathematics.

Already published

Basic Category Theory

TOM LEINSTER
University of Edinburgh

CAMBRIDGE
UNIVERSITY PRESS

University Printing House, Cambridge CB2 8BS, United Kingdom

One Liberty Plaza, 20th Floor, New York, NY 10006, USA

477 Williamstown Road, Port Melbourne, VIC 3207, Australia

314-321, 3rd Floor, Plot 3, Splendor Forum, Jasola District Centre, New Delhi - 110025, India

79 Anson Road, #06-04/06, Singapore 079906

Cambridge University Press is part of the University of Cambridge.

It furthers the University's mission by disseminating knowledge in the pursuit of education, learning and research at the highest international levels of excellence.

www.cambridge.org
Information on this title: www.cambridge.org/9781107044241

© Tom Leinster 2014

First published 2014
Reprinted 2017

A catalogue record for this publication is available from the British Library

ISBN 978-1-107-04424-1 Hardback

Contents

Note to the reader

This is not a sophisticated text. In writing it, I have assumed no more mathematical knowledge than might be acquired from an undergraduate degree at an ordinary British university, and I have not assumed that you are used to learning mathematics by reading a book rather than attending lectures. Furthermore, the list of topics covered is deliberately short, omitting all but the most fundamental parts of category theory. A 'further reading' section points to suitable follow-on texts.

There are two things that every reader should know about this book. One concerns the examples, and the other is about the exercises.

Each new concept is illustrated with a generous supply of examples, but it is not necessary to understand them all. In courses I have taught based on earlier versions of this text, probably no student has had the background to understand every example. All that matters is to understand enough examples that you can connect the new concepts with mathematics that you already know.

As for the exercises, I join every other textbook author in exhorting you to do them; but there is a further important point. In subjects such as number theory and combinatorics, some questions are simple to state but extremely hard to answer. Basic category theory is not like that. To understand the question is very nearly to know the answer. In most of the exercises, there is only one possible way to proceed. So, if you are stuck on an exercise, a likely remedy is to go back through each term in the question and make sure that you understand it *in full*. Take your time. Understanding, rather than problem solving, is the main challenge of learning category theory.

Citations such as Mac Lane (1971) refer to the sources listed in 'Further reading'.

This book developed out of master's-level courses taught several times at the University of Glasgow and, before that, at the University of Cambridge. In turn, the Cambridge version was based on Part III courses taught for many years by Martin Hyland and Peter Johnstone. Although this text is significantly different from any of their courses, I am conscious that certain exercises, lines of development and even turns of phrase have persisted through that long evolution. I would like to record my indebtedness to them, as well as my thanks to François Petit, my past students, the anonymous reviewers, and the staff of Cambridge University Press.

Introduction

Category theory takes a bird's eye view of mathematics. From high in the sky, details become invisible, but we can spot patterns that were impossible to detect from ground level. How is the lowest common multiple of two numbers like the direct sum of two vector spaces? What do discrete topological spaces, free groups, and fields of fractions have in common? We will discover answers to these and many similar questions, seeing patterns in mathematics that you may never have seen before.

The most important concept in this book is that of *universal property*. The further you go in mathematics, especially pure mathematics, the more universal properties you will meet. We will spend most of our time studying different manifestations of this concept.

Like all branches of mathematics, category theory has its own special vocabulary, which we will meet as we go along. But since the idea of universal property is so important, I will use this introduction to explain it with no jargon at all, by means of examples.

Our first example of a universal property is very simple.

Example 0.1   Let 1 denote a set with one element. (It does not matter what this element is called.) Then 1 has the following property:

for all sets X, there exists a unique map from X to 1.

(In this context, the words 'map', 'mapping' and 'function' all mean the same thing.)

Indeed, let X be a set. There *exists* a map $X \to 1$, because we can define f: $X \to 1$ by taking $f(x)$ to be the single element of 1 for each $x \in X$. This is the *unique* map $X \to 1$, because there is no choice in the matter: any map $X \to 1$ must send each element of X to the single element of 1.

Phrases of the form 'there exists a unique such-and-such satisfying some

1

condition' are common in category theory. The phrase means that there is one and only one such-and-such satisfying the condition. To prove the existence part, we have to show that there is at least one. To prove the uniqueness part, we have to show that there is at most one; in other words, any two such-and-suches satisfying the condition are equal.

Properties such as this are called 'universal' because they state how the object being described (in this case, the set 1) relates to the entire universe in which it lives (in this case, the universe of sets). The property begins with the words '*for all* sets X', and therefore says something about the relationship between 1 and *every* set X: namely, that there is a unique map from X to 1.

Example 0.2   This example involves rings, which in this book are always taken to have a multiplicative identity, called 1. Similarly, homomorphisms of rings are understood to preserve multiplicative identities.

The ring $\mathbb{Z}$ has the following property: for all rings R, there exists a unique homomorphism $\mathbb{Z} \to R$.

To prove existence, let R be a ring. Define a function $\phi \colon \mathbb{Z} \to R$ by

$$
\phi(n) = \begin{cases} \underbrace{1 + \cdots + 1}_{n} & \text{if } n > 0, \\ 0 & \text{if } n = 0, \\ -\phi(-n) & \text{if } n < 0 \end{cases}
$$

($n \in \mathbb{Z}$). A series of elementary checks confirms that ϕ is a homomorphism.

To prove uniqueness, let R be a ring and let $\psi \colon \mathbb{Z} \to R$ be a homomorphism. We show that ψ is equal to the homomorphism ϕ just defined. Since homomorphisms preserve multiplicative identities, $\psi(1) = 1$. Since homomorphisms preserve addition,

$$
\psi(n) = \psi(\underbrace{1 + \cdots + 1}_{n}) = \underbrace{\psi(1) + \cdots + \psi(1)}_{n} = \underbrace{1 + \cdots + 1}_{n} = \phi(n)
$$

for all $n > 0$. Since homomorphisms preserve zero, $\psi(0) = 0 = \phi(0)$. Finally, since homomorphisms preserve negatives, $\psi(n) = -\psi(-n) = -\phi(-n) = \phi(n)$ whenever $n < 0$.

Crucially, there can be essentially only *one* object satisfying a given universal property. The word 'essentially' means that two objects satisfying the same universal property need not literally be equal, but they are always isomorphic. For example:

Lemma 0.3   *Let A be a ring with the following property: for all rings R, there exists a unique homomorphism $A \to R$. Then $A \cong \mathbb{Z}$.*

Proof Let us call a ring with this property 'initial'. We are given that A is initial, and we proved in Example 0.2 that $\mathbb{Z}$ is initial.

Since A is initial, there is a unique homomorphism $\phi\colon A \to \mathbb{Z}$. Since $\mathbb{Z}$ is initial, there is a unique homomorphism $\phi'\colon \mathbb{Z} \to A$. Now $\phi' \circ \phi\colon A \to A$ is a homomorphism, but so too is the identity map $1_A\colon A \to A$; hence, since A is initial, $\phi' \circ \phi = 1_A$. (This follows from the uniqueness part of initiality, taking 'R' to be A.) Similarly, $\phi \circ \phi' = 1_{\mathbb{Z}}$. So ϕ and ϕ' are mutually inverse, and therefore define an isomorphism between A and $\mathbb{Z}$. □

This proof has very little to do with rings. It really belongs at a higher level of generality. To properly understand this, and to convey more fully the idea of universal property, it will help to consider some more complex examples.

Example 0.4 Let V be a vector space with a basis $(v_s)_{s\in S}$. (For example, if V is finite-dimensional then we might take $S = \{1, \dots, n\}$.) If W is another vector space, we can specify a linear map from V to W simply by saying where the basis elements go. Thus, for any W, there is a natural one-to-one correspondence between

> linear maps $V \to W$

and

> functions $S \to W$.

This is because any function defined on the basis elements extends uniquely to a linear map on V.

Let us rephrase this last statement. Define a function $i\colon S \to V$ by $i(s) = v_s$ ($s \in S$). Then V together with i has the following universal property:

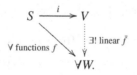

This diagram means that for all vector spaces W and all functions $f\colon S \to W$, there exists a unique linear map $\bar{f}\colon V \to W$ such that $\bar{f} \circ i = f$. The symbol $\forall$ means 'for all', and the symbols $\exists!$ mean 'there exists a unique'.

Another way to say '$\bar{f} \circ i = f$' is '$\bar{f}(v_s) = f(s)$ for all $s \in S$'. So, the diagram asserts that every function f defined on the basis elements extends uniquely to a linear map $\bar{f}$ defined on the whole of V. In other words still, the function

$$\{\text{linear maps } V \to W\} \to \{\text{functions } S \to W\}$$
$$\bar{f} \mapsto \bar{f} \circ i$$

is bijective.

Example 0.5   Given a set S, we can build a topological space $D(S)$ by equipping S with the **discrete topology**: all subsets are open. With this topology, *any* map from S to a space X is continuous.

Again, let us rephrase this. Define a function $i\colon S \to D(S)$ by $i(s) = s$ ($s \in S$). Then $D(S)$ together with i has the following universal property:

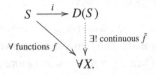

In other words, for all topological spaces X and all functions $f\colon S \to X$, there exists a unique continuous map $\bar{f}\colon D(S) \to X$ such that $\bar{f} \circ i = f$. The continuous map $\bar{f}$ is the same thing as the function f, except that we are regarding it as a continuous map between topological spaces rather than a mere function between sets.

You may feel that this universal property is almost too trivial to mean anything. But if we change the definition of $D(S)$ – say from the discrete to the indiscrete topology, in which the only open sets are $\emptyset$ and S – then the property becomes false. So this property really does say something about the discrete topology. What it says is that all maps out of a discrete space are continuous.

Indeed, given S, the universal property determines $D(S)$ and i uniquely (or rather, uniquely up to isomorphism; but who could want more?). The proof of this is similar to that of Lemma 0.3 above and Lemma 0.7 below.

Example 0.6   Given vector spaces U, V and W, a **bilinear map** $f\colon U \times V \to W$ is a function f that is linear in each variable:

$$f(u, v_1 + \lambda v_2) = f(u, v_1) + \lambda f(u, v_2),$$
$$f(u_1 + \lambda u_2, v) = f(u_1, v) + \lambda f(u_2, v)$$

for all $u, u_1, u_2 \in U$, $v, v_1, v_2 \in V$, and scalars λ. A good example is the scalar product (dot product), which is a bilinear map

$$\mathbb{R}^n \times \mathbb{R}^n \to \mathbb{R}$$
$$(\mathbf{u}, \mathbf{v}) \mapsto \mathbf{u}.\mathbf{v}$$

of real vector spaces. The vector product (cross product) $\mathbb{R}^3 \times \mathbb{R}^3 \to \mathbb{R}^3$ is also bilinear.

Let U and V be vector spaces. It is a fact that there is a 'universal bilinear map out of $U \times V'$. In other words, there exist a certain vector space T and a certain bilinear map $b \colon U \times V \to T$ with the following universal property:

$$
\begin{array}{ccc}
U \times V & \xrightarrow{\ b\ } & T \\
& \diagdown{\scriptstyle \forall\ \text{bilinear}\ f} & \ \Big\downarrow{\scriptstyle \exists!\ \text{linear}\ \bar{f}} \\
& & \forall W.
\end{array}
\tag{0.1}
$$

Roughly speaking, this property says that bilinear maps out of $U \times V$ correspond one-to-one with linear maps out of T.

Even without knowing that such a T and b exist, we can immediately prove that this universal property determines T and b uniquely up to isomorphism. The proof is essentially the same as that of Lemma 0.3, but looks more complicated because of the more complicated universal property.

Lemma 0.7 *Let U and V be vector spaces. Suppose that $b \colon U \times V \to T$ and $b' \colon U \times V \to T'$ are both universal bilinear maps out of $U \times V$. Then $T \cong T'$. More precisely, there exists a unique isomorphism $j \colon T \to T'$ such that $j \circ b = b'$.*

In the proof that follows, it does not actually matter what 'bilinear', 'linear' or even 'vector space' mean. The hard part is getting the logic straight. That done, you should be able to see that there is really only one possible proof. For instance, to use the universality of b, we will have to choose some bilinear map f out of $U \times V$. There are only two in sight, b and b', and we use each in the appropriate place.

Proof In diagram (0.1), take $\left(U \times V \xrightarrow{\ f\ } W \right)$ to be $\left(U \times V \xrightarrow{\ b'\ } T' \right)$. This gives a linear map $j \colon T \to T'$ satisfying $j \circ b = b'$. Similarly, using the universality of b', we obtain a linear map $j' \colon T' \to T$ satisfying $j' \circ b' = b$:

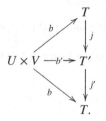

Now $j' \circ j \colon T \to T$ is a linear map satisfying $(j' \circ j) \circ b = b$; but also, the identity map $1_T \colon T \to T$ is linear and satisfies $1_T \circ b = b$. So by the uniqueness part of the universal property of b, we have $j' \circ j = 1_T$. (Here we took the 'f' of (0.1) to be b.) Similarly, $j \circ j' = 1_{T'}$. So j is an isomorphism. $\qquad\square$

In Example 0.6, it was stated that given vector spaces U and V, there exists a pair (T, b) with the universal property of (0.1). We just proved that there is essentially only one such pair (T, b). The vector space T is called the **tensor product** of U and V, and is written as $U \otimes V$. Tensor products are very important in algebra. They reduce the study of bilinear maps to the study of linear maps, since a bilinear map out of $U \times V$ is really the same thing as a linear map out of $U \otimes V$.

However, tensor products will not be important in this book. The real lesson for us is that it is safe to speak of *the* tensor product, not just *a* tensor product, and the reason for that is Lemma 0.7. This is a general point that applies to anything satisfying a universal property.

Once you know a universal property of an object, it often does no harm to forget how it was constructed. For instance, if you look through a pile of algebra books, you will find several different ways of constructing the tensor product of two vector spaces. But once you have proved that the tensor product satisfies the universal property, you can forget the construction. The universal property tells you all you need to know, because it determines the object uniquely up to isomorphism.

Example 0.8 Let $\theta \colon G \to H$ be a homomorphism of groups. Associated with θ is a diagram

$$\ker(\theta) \overset{\iota}{\hookrightarrow} G \underset{\varepsilon}{\overset{\theta}{\rightrightarrows}} H, \qquad (0.2)$$

where ι is the inclusion of $\ker(\theta)$ into G and ε is the trivial homomorphism. 'Inclusion' means that $\iota(x) = x$ for all $x \in \ker(\theta)$, and 'trivial' means that $\varepsilon(g) = 1$ for all $g \in G$. The symbol $\hookrightarrow$ is often used for inclusions; it is a combination of a subset symbol $\subset$ and an arrow.

The map ι into G satisfies $\theta \circ \iota = \varepsilon \circ \iota$, and is universal as such. Exercise 0.11 asks you to make this precise.

Here is our final example of a universal property.

Example 0.9 Take a topological space covered by two open subsets: $X = U \cup V$. The diagram

$$\begin{array}{ccc} U \cap V & \overset{i}{\hookrightarrow} & U \\ \scriptstyle{j} \big\downarrow & & \big\downarrow \scriptstyle{j'} \\ V & \underset{i'}{\hookrightarrow} & X \end{array}$$

of inclusion maps has a universal property in the world of topological spaces and continuous maps, as follows:

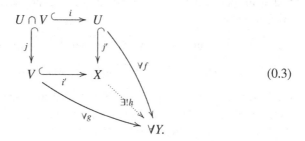

$$(0.3)$$

The diagram means that given Y, f and g such that $f \circ i = g \circ j$, there is exactly one continuous map $h \colon X \to Y$ such that $h \circ j' = f$ and $h \circ i' = g$.

Under favourable conditions, the induced diagram

$$
\begin{array}{ccc}
\pi_1(U \cap V) & \xrightarrow{\;i_*\;} & \pi_1(U) \\
{\scriptstyle j_*}\big\downarrow & & \big\downarrow{\scriptstyle j'_*} \\
\pi_1(V) & \xrightarrow[\;i'_*\;]{} & \pi_1(X)
\end{array}
$$

of fundamental groups has the same property in the world of groups and group homomorphisms. This is *van Kampen's theorem*. In fact, van Kampen stated his theorem in a much more complicated way. Stating it transparently requires some categorical language, but he was working in the 1930s, before category theory had been born.

You have now seen several examples of universal properties. As this book progresses, we will develop different ways of talking about them. Once we have set up the basic vocabulary of categories and functors, we will study *adjoint functors*, then *representable functors*, then *limits*. Each of these provides an approach to universal properties, and each places the idea in a different light. For instance, Examples 0.4 and 0.5 can most readily be described in terms of adjoint functors, Example 0.6 via representable functors, and Examples 0.1, 0.2, 0.8 and 0.9 in terms of limits.

Exercises

0.10 Let S be a set. The **indiscrete** topological space $I(S)$ is the space whose set of points is S and whose only open subsets are $\emptyset$ and S itself. Imitating Example 0.5, find a universal property satisfied by the space $I(S)$.

0.11 Fix a group homomorphism $\theta \colon G \to H$. Find a universal property satisfied by the pair $(\ker(\theta), \iota)$ of diagram (0.2). (This property can – indeed, must – make reference to θ.)

0.12 Verify the universal property shown in diagram (0.3).

0.13 Denote by $\mathbb{Z}[x]$ the polynomial ring over $\mathbb{Z}$ in one variable.

(a) Prove that for all rings R and all $r \in R$, there exists a unique ring homomorphism $\phi \colon \mathbb{Z}[x] \to R$ such that $\phi(x) = r$.
(b) Let A be a ring and $a \in A$. Suppose that for all rings R and all $r \in R$, there exists a unique ring homomorphism $\phi \colon A \to R$ such that $\phi(a) = r$. Prove that there is a unique isomorphism $\iota \colon \mathbb{Z}[x] \to A$ such that $\iota(x) = a$.

0.14 Let X and Y be vector spaces.

(a) For the purposes of this exercise only, a *cone* is a triple (V, f_1, f_2) consisting of a vector space V, a linear map $f_1 \colon V \to X$, and a linear map $f_2 \colon V \to Y$. Find a cone (P, p_1, p_2) with the following property: for all cones (V, f_1, f_2), there exists a unique linear map $f \colon V \to P$ such that $p_1 \circ f = f_1$ and $p_2 \circ f = f_2$.
(b) Prove that there is essentially only one cone with the property stated in (a). That is, prove that if (P, p_1, p_2) and (P', p_1', p_2') both have this property then there is an isomorphism $i \colon P \to P'$ such that $p_1' \circ i = p_1$ and $p_2' \circ i = p_2$.
(c) For the purposes of this exercise only, a *cocone* is a triple (V, f_1, f_2) consisting of a vector space V, a linear map $f_1 \colon X \to V$, and a linear map $f_2 \colon Y \to V$. Find a cocone (Q, q_1, q_2) with the following property: for all cocones (V, f_1, f_2), there exists a unique linear map $f \colon Q \to V$ such that $f \circ q_1 = f_1$ and $f \circ q_2 = f_2$.
(d) Prove that there is essentially only one cocone with the property stated in (c), in a sense that you should make precise.

1

Categories, functors and natural transformations

A category is a system of related objects. The objects do not live in isolation: there is some notion of map between objects, binding them together.

Typical examples of what 'object' might mean are 'group' and 'topological space', and typical examples of what 'map' might mean are 'homomorphism' and 'continuous map', respectively. We will see many examples, and we will also learn that some categories have a very different flavour from the two just mentioned. In fact, the 'maps' of category theory need not be anything like maps in the sense that you are most likely to be familiar with.

Categories are *themselves* mathematical objects, and with that in mind, it is unsurprising that there is a good notion of 'map between categories'. Such maps are called functors. More surprising, perhaps, is the existence of a third level: we can talk about maps between *functors*, which are called natural transformations. These, then, are maps between maps between categories.

In fact, it was the desire to formalize the notion of natural transformation that led to the birth of category theory. By the early 1940s, researchers in algebraic topology had started to use the phrase 'natural transformation', but only in an informal way. Two mathematicians, Samuel Eilenberg and Saunders Mac Lane, saw that a precise definition was needed. But before they could define natural transformation, they had to define functor; and before they could define functor, they had to define category. And so the subject was born.

Nowadays, the uses of category theory have spread far beyond algebraic topology. Its tentacles extend into most parts of pure mathematics. They also reach some parts of applied mathematics; perhaps most notably, category theory has become a standard tool in certain parts of computer science. Applied mathematics is more than just applied differential equations!

9

1.1 Categories

Definition 1.1.1   A **category** $\mathscr{A}$ consists of:

- a collection $\mathrm{ob}(\mathscr{A})$ of **objects**;
- for each $A, B \in \mathrm{ob}(\mathscr{A})$, a collection $\mathscr{A}(A, B)$ of **maps** or **arrows** or **morphisms** from A to B;
- for each $A, B, C \in \mathrm{ob}(\mathscr{A})$, a function

$$\mathscr{A}(B, C) \times \mathscr{A}(A, B) \quad \to \quad \mathscr{A}(A, C)$$
$$(g, f) \quad \mapsto \quad g \circ f,$$

called **composition**;

- for each $A \in \mathrm{ob}(\mathscr{A})$, an element 1_A of $\mathscr{A}(A, A)$, called the **identity** on A,

satisfying the following axioms:

- **associativity**: for each $f \in \mathscr{A}(A, B)$, $g \in \mathscr{A}(B, C)$ and $h \in \mathscr{A}(C, D)$, we have $(h \circ g) \circ f = h \circ (g \circ f)$;
- **identity laws**: for each $f \in \mathscr{A}(A, B)$, we have $f \circ 1_A = f = 1_B \circ f$.

Remarks 1.1.2   (a) We often write:

$$A \in \mathscr{A} \quad \text{to mean} \quad A \in \mathrm{ob}(\mathscr{A});$$
$$f \colon A \to B \text{ or } A \xrightarrow{f} B \quad \text{to mean} \quad f \in \mathscr{A}(A, B);$$
$$gf \quad \text{to mean} \quad g \circ f.$$

People also write $\mathscr{A}(A, B)$ as $\mathrm{Hom}_{\mathscr{A}}(A, B)$ or $\mathrm{Hom}(A, B)$. The notation 'Hom' stands for homomorphism, from one of the earliest examples of a category.

(b) The definition of category is set up so that in general, from each string

$$A_0 \xrightarrow{f_1} A_1 \xrightarrow{f_2} \cdots \xrightarrow{f_n} A_n$$

of maps in $\mathscr{A}$, it is possible to construct exactly one map

$$A_0 \to A_n$$

(namely, $f_n f_{n-1} \cdots f_1$). If we are given extra information then we may be able to construct other maps $A_0 \to A_n$; for instance, if we happen to know that $A_{n-1} = A_n$, then $f_{n-1} f_{n-2} \cdots f_1$ is another such map. But we are speaking here of the *general* situation, in the absence of extra information.

For example, a string like this with $n = 4$ gives rise to maps

$$A_0 \underset{(f_4(1_{A_3} f_3))((f_2 f_1) 1_{A_0})}{\overset{((f_4 f_3) f_2) f_1}{\rightrightarrows}} A_4,$$

but the axioms imply that they are equal. It is safe to omit the brackets and write both as $f_4f_3f_2f_1$.

Here it is intended that $n \geq 0$. In the case $n = 0$, the statement is that for each object A_0 of a category, it is possible to construct exactly one map $A_0 \to A_0$ (namely, the identity 1_{A_0}). An identity map can be thought of as a zero-fold composite, in much the same way that the number 1 can be thought of as the product of zero numbers.

(c) We often speak of **commutative diagrams**. For instance, given objects and maps

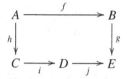

in a category, we say that the diagram **commutes** if $gf = jih$. Generally, a diagram is said to commute if whenever there are two paths from an object X to an object Y, the map from X to Y obtained by composing along one path is equal to the map obtained by composing along the other.

(d) The slightly vague word 'collection' means *roughly* the same as 'set', although if you know about such things, it is better to interpret it as meaning 'class'. We come back to this in Chapter 3.

(e) If $f \in \mathscr{A}(A, B)$, we call A the **domain** and B the **codomain** of f. Every map in every category has a definite domain and a definite codomain. (If you believe it makes sense to form the intersection of an arbitrary pair of abstract sets, you should add to the definition of category the condition that $\mathscr{A}(A, B) \cap \mathscr{A}(A', B') = \emptyset$ unless $A = A'$ and $B = B'$.)

Examples 1.1.3 (Categories of mathematical structures) (a) There is a category **Set** described as follows. Its objects are sets. Given sets A and B, a map from A to B in the category **Set** is exactly what is ordinarily called a map (or mapping, or function) from A to B. Composition in the category is ordinary composition of functions, and the identity maps are again what you would expect.

In situations such as this, we often do not bother to specify the composition and identities. We write 'the category of sets and functions', leaving the reader to guess the rest. In fact, we usually go further and call it just 'the category of sets'.

(b) There is a category **Grp** of groups, whose objects are groups and whose maps are group homomorphisms.

(c) Similarly, there is a category **Ring** of rings and ring homomorphisms.

(d) For each field k, there is a category **Vect**$_k$ of vector spaces over k and linear maps between them.

(e) There is a category **Top** of topological spaces and continuous maps.

This chapter is mostly about the interaction *between* categories, rather than what goes on *inside* them. We will, however, need the following definition.

Definition 1.1.4 A map $f: A \to B$ in a category $\mathscr{A}$ is an **isomorphism** if there exists a map $g: B \to A$ in $\mathscr{A}$ such that $gf = 1_A$ and $fg = 1_B$.

In the situation of Definition 1.1.4, we call g the **inverse** of f and write $g = f^{-1}$. (The word 'the' is justified by Exercise 1.1.13.) If there exists an isomorphism from A to B, we say that A and B are **isomorphic** and write $A \cong B$.

Example 1.1.5 The isomorphisms in **Set** are exactly the bijections. This statement is not quite a logical triviality. It amounts to the assertion that a function has a two-sided inverse if and only if it is injective and surjective.

Example 1.1.6 The isomorphisms in **Grp** are exactly the isomorphisms of groups. Again, this is not quite trivial, at least if you were taught that the definition of group isomorphism is 'bijective homomorphism'. In order to show that this is equivalent to being an isomorphism in **Grp**, you have to prove that the inverse of a bijective homomorphism is also a homomorphism.

Similarly, the isomorphisms in **Ring** are exactly the isomorphisms of rings.

Example 1.1.7 The isomorphisms in **Top** are exactly the homeomorphisms. Note that, in contrast to the situation in **Grp** and **Ring**, a bijective map in **Top** is not necessarily an isomorphism. A classic example is the map

$$[0, 1) \to \{z \in \mathbb{C} \mid |z| = 1\}$$
$$t \mapsto e^{2\pi i t},$$

which is a continuous bijection but not a homeomorphism.

The examples of categories mentioned so far are important, but could give a false impression. In each of them, the objects of the category are sets with structure (such as a group structure, a topology, or, in the case of **Set**, no structure at all). The maps are the functions preserving the structure, in the appropriate sense. And in each of them, there is a clear sense of what the elements of a given object are.

However, not all categories are like this. In general, the objects of a category are not 'sets equipped with extra stuff'. Thus, in a general category, it does not make sense to talk about the 'elements' of an object. (At least, it does not make sense in an immediately obvious way; we return to this in Definition 4.1.25.) Similarly, in a general category, the maps need not be mappings or functions in the usual sense. So:

The objects of a category need not be remotely like sets.

The maps in a category need not be remotely like functions.

The next few examples illustrate these points. They also show that, contrary to the impression that might have been given so far, categories need not be enormous. Some categories are small, manageable structures in their own right, as we now see.

Examples 1.1.8 (Categories as mathematical structures)   (a) A category can be specified by saying directly what its objects, maps, composition and identities are. For example, there is a category $\emptyset$ with no objects or maps at all. There is a category **1** with one object and only the identity map. It can be drawn like this:

$$\bullet$$

(Since every object is required to have an identity map on it, we usually do not bother to draw the identities.) There is another category that can be drawn as

$$\bullet \to \bullet \qquad \text{or} \qquad A \xrightarrow{f} B,$$

with two objects and one non-identity map, from the first object to the second. (Composition is defined in the only possible way.) To reiterate the points made above, it is not obvious what an 'element' of A or B would be, or how one could regard f as a 'function' of any sort.

It is easy to make up more complicated examples. For instance, here are three more categories:

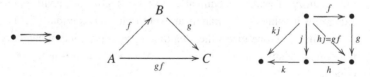

(b) Some categories contain no maps at all apart from identities (which, as categories, they are obliged to have). These are called **discrete** categories. A discrete category amounts to just a class of objects. More poetically, a category is a collection of objects related to one another to a greater or lesser degree; a discrete category is the extreme case in which each object is totally isolated from its companions.

(c) A group is essentially the same thing as a category that has only one object and in which all the maps are isomorphisms.

To understand this, first consider a category $\mathscr{A}$ with just one object. It is not important what letter or symbol we use to denote the object; let us call it A. Then $\mathscr{A}$ consists of a set (or class) $\mathscr{A}(A, A)$, an associative composition function

$$\circ : \mathscr{A}(A, A) \times \mathscr{A}(A, A) \to \mathscr{A}(A, A),$$

and a two-sided unit $1_A \in \mathscr{A}(A, A)$. This would make $\mathscr{A}(A, A)$ into a group, except that we have not mentioned inverses. However, to say that every map in $\mathscr{A}$ is an isomorphism is exactly to say that every element of $\mathscr{A}(A, A)$ has an inverse with respect to $\circ$.

If we write G for the group $\mathscr{A}(A, A)$, then the situation is this:

category $\mathscr{A}$ with single object A	corresponding group G
maps in $\mathscr{A}$	elements of G
$\circ$ in $\mathscr{A}$	$\cdot$ in G
1_A	$1 \in G$

The category $\mathscr{A}$ looks something like this:

The arrows represent different maps $A \to A$, that is, different elements of the group G.

What the object of $\mathscr{A}$ is called makes no difference. It matters exactly as much as whether we choose x or y or t to denote some variable in an algebra problem, which is to say, not at all. Later we will define 'equivalence' of categories, which will enable us to make a precise statement: the category of groups is equivalent to the category of (small) one-object categories in which every map is an isomorphism (Example 3.2.11).

The first time one meets the idea that a group is a kind of category, it is tempting to dismiss it as a coincidence or a trick. But it is not; there is real content.

To see this, suppose that your education had been shuffled and that you already knew about categories before being taught about groups. In your first group theory class, the lecturer declares that a group is supposed to be the system of all symmetries of an object. A symmetry of an object X, she says, is a way of mapping X to itself in a reversible or invertible manner. At this point, you realize that she is talking about a very special type of category. In general, a category is a system consisting of *all* the mappings (not usually just the invertible ones) between *many* objects (not usually just one). So a group is just a category with the special properties that all the maps are invertible and there is only one object.

(d) The inverses played no essential part in the previous example, suggesting that it is worth thinking about 'groups without inverses'. These are called monoids.

Formally, a **monoid** is a set equipped with an associative binary operation and a two-sided unit element. Groups describe the reversible transformations, or symmetries, that can be applied to an object; monoids describe the not-necessarily-reversible transformations. For instance, given any set X, there is a group consisting of all bijections $X \to X$, and there is a monoid consisting of all functions $X \to X$. In both cases, the binary operation is composition and the unit is the identity function on X. Another example of a monoid is the set $\mathbb{N} = \{0, 1, 2, \ldots\}$ of natural numbers, with $+$ as the operation and 0 as the unit. Alternatively, we could take the set $\mathbb{N}$ with $\cdot$ as the operation and 1 as the unit.

A category with one object is essentially the same thing as a monoid, by the same argument as for groups. This is stated formally in Example 3.2.11.

(e) A **preorder** is a reflexive transitive binary relation. A **preordered set** $(S, \leq)$ is a set S together with a preorder $\leq$ on it. Examples: $S = \mathbb{R}$ and $\leq$ has its usual meaning; S is the set of subsets of $\{1, \ldots, 10\}$ and $\leq$ is $\subseteq$ (inclusion); $S = \mathbb{Z}$ and $a \leq b$ means that a divides b.

A preordered set can be regarded as a category $\mathscr{A}$ in which, for each $A, B \in \mathscr{A}$, there is at most one map from A to B. To see this, consider a category $\mathscr{A}$ with this property. It is not important what letter we use to denote the unique map from an object A to an object B; all we need to record is which pairs (A, B) of objects have the property that a map $A \to B$ does exist. Let us write $A \leq B$ to mean that there exists a map $A \to B$.

Since $\mathscr{A}$ is a category, and categories have composition, if $A \leq B \leq C$ then $A \leq C$. Since categories also have identities, $A \leq A$ for all A. The associativity and identity axioms are automatic. So, $\mathscr{A}$ amounts to a collection of objects equipped with a transitive reflexive binary relation, that is, a preorder. One can think of the unique map $A \to B$ as the statement or assertion that $A \leq B$.

An **order** on a set is a preorder $\leq$ with the property that if $A \leq B$ and $B \leq A$ then $A = B$. (Equivalently, if $A \cong B$ in the corresponding category then $A = B$.) Ordered sets are also called **partially ordered sets** or **posets**.

An example of a preorder that is not an order is the divisibility relation |
on $\mathbb{Z}$: for there we have $2 \mid -2$ and $-2 \mid 2$ but $2 \neq -2$.

Here are two ways of constructing new categories from old.

Construction 1.1.9   Every category $\mathscr{A}$ has an **opposite** or **dual** category
$\mathscr{A}^{\mathrm{op}}$, defined by reversing the arrows. Formally, $\mathrm{ob}(\mathscr{A}^{\mathrm{op}}) = \mathrm{ob}(\mathscr{A})$ and
$\mathscr{A}^{\mathrm{op}}(B, A) = \mathscr{A}(A, B)$ for all objects A and B. Identities in $\mathscr{A}^{\mathrm{op}}$ are the
same as in $\mathscr{A}$. Composition in $\mathscr{A}^{\mathrm{op}}$ is the same as in $\mathscr{A}$, but with the arguments reversed. To spell this out: if $A \xrightarrow{f} B \xrightarrow{g} C$ are maps in $\mathscr{A}^{\mathrm{op}}$ then
$A \xleftarrow{f} B \xleftarrow{g} C$ are maps in $\mathscr{A}$; these give rise to a map $A \xleftarrow{f \circ g} C$ in $\mathscr{A}$, and
the composite of the original pair of maps is the corresponding map $A \to C$ in
$\mathscr{A}^{\mathrm{op}}$.

So, arrows $A \to B$ in $\mathscr{A}$ correspond to arrows $B \to A$ in $\mathscr{A}^{\mathrm{op}}$. According
to the definition above, if $f \colon A \to B$ is an arrow in $\mathscr{A}$ then the corresponding
arrow $B \to A$ in $\mathscr{A}^{\mathrm{op}}$ is also called f. Some people prefer to give it a different
name, such as f^{op}.

Remark 1.1.10   The **principle of duality** is fundamental to category theory.
Informally, it states that every categorical definition, theorem and proof has
a **dual**, obtained by reversing all the arrows. Invoking the principle of duality can save work: given any theorem, reversing the arrows throughout its
statement and proof produces a dual theorem. Numerous examples of duality
appear throughout this book.

Construction 1.1.11   Given categories $\mathscr{A}$ and $\mathscr{B}$, there is a **product category** $\mathscr{A} \times \mathscr{B}$, in which

$$\mathrm{ob}(\mathscr{A} \times \mathscr{B}) = \mathrm{ob}(\mathscr{A}) \times \mathrm{ob}(\mathscr{B}),$$
$$(\mathscr{A} \times \mathscr{B})((A, B), (A', B')) = \mathscr{A}(A, A') \times \mathscr{B}(B, B').$$

Put another way, an object of the product category $\mathscr{A} \times \mathscr{B}$ is a pair (A, B) where
$A \in \mathscr{A}$ and $B \in \mathscr{B}$. A map $(A, B) \to (A', B')$ in $\mathscr{A} \times \mathscr{B}$ is a pair (f, g) where
$f \colon A \to A'$ in $\mathscr{A}$ and $g \colon B \to B'$ in $\mathscr{B}$. For the definitions of composition and
identities in $\mathscr{A} \times \mathscr{B}$, see Exercise 1.1.14.

Exercises

1.1.12   Find three examples of categories not mentioned above.

1.1.13   Show that a map in a category can have at most one inverse. That is,
given a map $f \colon A \to B$, show that there is at most one map $g \colon B \to A$ such
that $gf = 1_A$ and $fg = 1_B$.

1.1.14 Let $\mathscr{A}$ and $\mathscr{B}$ be categories. Construction 1.1.11 defined the product category $\mathscr{A} \times \mathscr{B}$, except that the definitions of composition and identities in $\mathscr{A} \times \mathscr{B}$ were not given. There is only one sensible way to define them; write it down.

1.1.15 There is a category **Toph** whose objects are topological spaces and whose maps $X \to Y$ are homotopy classes of continuous maps from X to Y. What do you need to know about homotopy in order to prove that **Toph** is a category? What does it mean, in purely topological terms, for two objects of **Toph** to be isomorphic?

1.2 Functors

One of the lessons of category theory is that whenever we meet a new type of mathematical object, we should always ask whether there is a sensible notion of 'map' between such objects. We can ask this about categories themselves. The answer is yes, and a map between categories is called a functor.

Definition 1.2.1 Let $\mathscr{A}$ and $\mathscr{B}$ be categories. A **functor** $F: \mathscr{A} \to \mathscr{B}$ consists of:

- a function
$$\mathrm{ob}(\mathscr{A}) \to \mathrm{ob}(\mathscr{B}),$$
written as $A \mapsto F(A)$;
- for each $A, A' \in \mathscr{A}$, a function
$$\mathscr{A}(A, A') \to \mathscr{B}(F(A), F(A')),$$
written as $f \mapsto F(f)$,

satisfying the following axioms:

- $F(f' \circ f) = F(f') \circ F(f)$ whenever $A \xrightarrow{f} A' \xrightarrow{f'} A''$ in $\mathscr{A}$;
- $F(1_A) = 1_{F(A)}$ whenever $A \in \mathscr{A}$.

Remarks 1.2.2 (a) The definition of functor is set up so that from each string
$$A_0 \xrightarrow{f_1} \cdots \xrightarrow{f_n} A_n$$
of maps in $\mathscr{A}$ (with $n \geq 0$), it is possible to construct exactly one map
$$F(A_0) \to F(A_n)$$

in $\mathcal{B}$. For example, given maps

$$A_0 \xrightarrow{f_1} A_1 \xrightarrow{f_2} A_2 \xrightarrow{f_3} A_3 \xrightarrow{f_4} A_4$$

in $\mathcal{A}$, we can construct maps

$$F(A_0) \xrightarrow[\quad F(1_{A_4})F(f_4)F(f_3f_2)F(f_1) \quad]{\quad F(f_4f_3)F(f_2f_1) \quad} F(A_4)$$

in $\mathcal{B}$, but the axioms imply that they are equal.

(b) We are familiar with the idea that structures and the structure-preserving maps between them form a category (such as **Grp**, **Ring**, etc.). In particular, this applies to categories and functors: there is a category **CAT** whose objects are categories and whose maps are functors.

One part of this statement is that functors can be composed. That is, given functors $\mathcal{A} \xrightarrow{F} \mathcal{B} \xrightarrow{G} \mathcal{C}$, there arises a new functor $\mathcal{A} \xrightarrow{G \circ F} \mathcal{C}$, defined in the obvious way. Another is that for every category $\mathcal{A}$, there is an identity functor $1_{\mathcal{A}} : \mathcal{A} \to \mathcal{A}$.

Examples 1.2.3 Perhaps the easiest examples of functors are the so-called **forgetful functors**. (This is an informal term, with no precise definition.) For instance:

(a) There is a functor $U \colon \mathbf{Grp} \to \mathbf{Set}$ defined as follows: if G is a group then $U(G)$ is the underlying set of G (that is, its set of elements), and if $f \colon G \to H$ is a group homomorphism then $U(f)$ is the function f itself. So U forgets the group structure of groups and forgets that group homomorphisms are homomorphisms.

(b) Similarly, there is a functor **Ring** $\to$ **Set** forgetting the ring structure on rings, and (for any field k) there is a functor $\mathbf{Vect}_k \to \mathbf{Set}$ forgetting the vector space structure on vector spaces.

(c) Forgetful functors do not have to forget *all* the structure. For example, let **Ab** be the category of abelian groups. There is a functor **Ring** $\to$ **Ab** that forgets the multiplicative structure, remembering just the underlying additive group. Or, let **Mon** be the category of monoids. There is a functor $U \colon \mathbf{Ring} \to \mathbf{Mon}$ that forgets the additive structure, remembering just the underlying multiplicative monoid. (That is, if R is a ring then $U(R)$ is the set R made into a monoid via $\cdot$ and 1.)

(d) There is an inclusion functor $U \colon \mathbf{Ab} \to \mathbf{Grp}$ defined by $U(A) = A$ for any abelian group A and $U(f) = f$ for any homomorphism f of abelian groups. It forgets that abelian groups are abelian.

The forgetful functors in examples (a)–(c) forget *structure* on the objects, but that of example (d) forgets a *property*. Nevertheless, it turns out to be convenient to use the same word, 'forgetful', in both situations.

Although forgetting is a trivial operation, there are situations in which it is powerful. For example, it is a theorem that the order of any finite field is a prime power. An important step in the proof is to simply forget that the field is a field, remembering only that it is a vector space over its subfield $\{0, 1, 1 + 1, 1 + 1 + 1, \ldots\}$.

Examples 1.2.4 **Free functors** are in some sense dual to forgetful functors (as we will see in the next chapter), although they are less elementary. Again, 'free functor' is an informal but useful term.

(a) Given any set S, one can build the **free group** $F(S)$ on S. This is a group containing S as a subset and with no further properties other than those it is forced to have, in a sense made precise in Section 2.1. Intuitively, the group $F(S)$ is obtained from the set S by adding just enough new elements that it becomes a group, but without imposing any equations other than those forced by the definition of group.

A little more precisely, the elements of $F(S)$ are formal expressions or **words** such as $x^{-4}yx^2zy^{-3}$ (where $x, y, z \in S$). Two such words are seen as equal if one can be obtained from the other by the usual cancellation rules, so that, for example, x^3xy, x^4y, and $x^2y^{-1}yx^2y$ all represent the same element of $F(S)$. To multiply two words, just write one followed by the other; for instance, $x^{-4}yx$ times xzy^{-3} is $x^{-4}yx^2zy^{-3}$.

This construction assigns to each set S a group $F(S)$. In fact, F is a functor: any map of sets $f: S \to S'$ gives rise to a homomorphism of groups $F(f): F(S) \to F(S')$. For instance, take the map of sets

$$f: \{w, x, y, z\} \to \{u, v\}$$

defined by $f(w) = f(x) = f(y) = u$ and $f(z) = v$. This gives rise to a homomorphism

$$F(f): F(\{w, x, y, z\}) \to F(\{u, v\}),$$

which maps $x^{-4}yx^2zy^{-3} \in F(\{w, x, y, z\})$ to

$$u^{-4}uu^2vu^{-3} = u^{-1}vu^{-3} \in F(\{u, v\}).$$

(b) Similarly, we can construct the free commutative ring $F(S)$ on a set S, giving a functor F from **Set** to the category **CRing** of commutative rings. In fact, $F(S)$ is something familiar, namely, the ring of polynomials over $\mathbb{Z}$ in commuting variables x_s ($s \in S$). (A polynomial is, after all, just a formal expression built from the variables using the ring operations $+$, $-$ and $\cdot$.) For example, if S is a two-element set then $F(S) \cong \mathbb{Z}[x, y]$.

(c) We can also construct the free vector space on a set. Fix a field k. The free functor $F \colon \mathbf{Set} \to \mathbf{Vect}_k$ is defined on objects by taking $F(S)$ to be a vector space with basis S. Any two such vector spaces are isomorphic; but it is perhaps not obvious that there is any such vector space at all, so we have to construct one. Loosely, $F(S)$ is the set of all formal k-linear combinations of elements of S, that is, expressions

$$\sum_{s \in S} \lambda_s s$$

where each λ_s is a scalar and there are only finitely many values of s such that $\lambda_s \neq 0$. (This restriction is imposed because one can only take *finite* sums in a vector space.) Elements of $F(S)$ can be added:

$$\sum_{s \in S} \lambda_s s + \sum_{s \in S} \mu_s s = \sum_{s \in S} (\lambda_s + \mu_s)s.$$

There is also a scalar multiplication on $F(S)$:

$$c \cdot \sum_{s \in S} \lambda_s s = \sum_{s \in S} (c\lambda_s)s$$

($c \in k$). In this way, $F(S)$ becomes a vector space.

To be completely precise and avoid talking about 'expressions', we can define $F(S)$ to be the set of all functions $\lambda \colon S \to k$ such that $\{s \in S \mid \lambda(s) \neq 0\}$ is finite. (Think of such a function λ as corresponding to the expression $\sum_{s \in S} \lambda(s)s$.) To define addition on $F(S)$, we must define for each $\lambda, \mu \in F(S)$ a sum $\lambda + \mu \in F(S)$; it is given by

$$(\lambda + \mu)(s) = \lambda(s) + \mu(s)$$

($s \in S$). Similarly, the scalar multiplication is given by $(c \cdot \lambda)(s) = c \cdot \lambda(s)$ ($c \in k$, $\lambda \in F(S)$, $s \in S$).

Rings and vector spaces have the special property that it is relatively easy to write down an explicit formula for the free functor. The case of groups is much more typical. For most types of algebraic structure, describing the free functor requires as much fussy work as it does for groups. We return to this point in Example 2.1.3 and Example 6.3.11 (where we see how to avoid the fussy work entirely).

Examples 1.2.5 (Functors in algebraic topology) Historically, some of the first examples of functors arose in algebraic topology. There, the strategy is to learn about a space by extracting data from it in some clever way, assembling that data into an algebraic structure, then studying the algebraic structure instead of the original space. Algebraic topology therefore involves many functors from categories of spaces to categories of algebras.

(a) Let **Top**$_*$ be the category of topological spaces equipped with a basepoint, together with the continuous basepoint-preserving maps. There is a functor $\pi_1 \colon$ **Top**$_* \to$ **Grp** assigning to each space X with basepoint x the fundamental group $\pi_1(X, x)$ of X at x. (Some texts use the simpler notation $\pi_1(X)$, ignoring the choice of basepoint. This is more or less safe if X is path-connected, but strictly speaking, the basepoint should always be specified.)

   That π_1 is a functor means that it not only assigns to each space-with-basepoint (X, x) a group $\pi_1(X, x)$, but also assigns to each basepoint-preserving continuous map

$$f \colon (X, x) \to (Y, y)$$

a homomorphism

$$\pi_1(f) \colon \pi_1(X, x) \to \pi_1(Y, y).$$

Usually $\pi_1(f)$ is written as f_*. The functoriality axioms say that $(g \circ f)_* = g_* \circ f_*$ and $(1_{(X,x)})_* = 1_{\pi_1(X,x)}$.

(b) For each $n \in \mathbb{N}$, there is a functor $H_n \colon$ **Top** $\to$ **Ab** assigning to a space its nth homology group (in any of several possible senses).

Example 1.2.6 Any system of polynomial equations such as

$$2x^2 + y^2 - 3z^2 = 1 \tag{1.1}$$

$$x^3 + x = y^2 \tag{1.2}$$

gives rise to a functor **CRing** $\to$ **Set**. Indeed, for each commutative ring A, let $F(A)$ be the set of triples $(x, y, z) \in A \times A \times A$ satisfying equations (1.1) and (1.2). Whenever $f \colon A \to B$ is a ring homomorphism and $(x, y, z) \in F(A)$, we have $(f(x), f(y), f(z)) \in F(B)$; so the map of rings $f \colon A \to B$ induces a map of sets $F(f) \colon F(A) \to F(B)$. This defines a functor $F \colon$ **CRing** $\to$ **Set**.

   In algebraic geometry, a **scheme** is a functor **CRing** $\to$ **Set** with certain properties. (This is not the most common way of phrasing the definition, but it is equivalent.) The functor F above is a simple example.

Example 1.2.7 Let G and H be monoids (or groups, if you prefer), regarded as one-object categories $\mathscr{G}$ and $\mathscr{H}$. A functor $F \colon \mathscr{G} \to \mathscr{H}$ must send the unique object of $\mathscr{G}$ to the unique object of $\mathscr{H}$, so it is determined by its effect on maps. Hence, the functor $F\colon \mathscr{G} \to \mathscr{H}$ amounts to a function $F\colon G \to H$ such that $F(g'g) = F(g')F(g)$ for all $g', g \in G$, and $F(1) = 1$. In other words, a functor $\mathscr{G} \to \mathscr{H}$ is just a homomorphism $G \to H$.

Example 1.2.8   Let G be a monoid, regarded as a one-object category $\mathscr{G}$. A functor $F\colon \mathscr{G} \to \mathbf{Set}$ consists of a set S (the value of F at the unique object of $\mathscr{G}$) together with, for each $g \in G$, a function $F(g)\colon S \to S$, satisfying the functoriality axioms. Writing $(F(g))(s) = g \cdot s$, we see that the functor F amounts to a set S together with a function

$$
\begin{aligned}
G \times S &\to S \\
(g, s) &\mapsto g \cdot s
\end{aligned}
$$

satisfying $(g'g) \cdot s = g' \cdot (g \cdot s)$ and $1 \cdot s = s$ for all $g, g' \in G$ and $s \in S$. In other words, a functor $\mathscr{G} \to \mathbf{Set}$ is a set equipped with a left action by G: a **left G-set**, for short.

Similarly, a functor $\mathscr{G} \to \mathbf{Vect}_k$ is exactly a k-linear representation of G, in the sense of representation theory. This can reasonably be taken as the *definition* of representation.

Example 1.2.9   When A and B are (pre)ordered sets, a functor between the corresponding categories is exactly an **order-preserving map**, that is, a function $f\colon A \to B$ such that $a \leq a' \implies f(a) \leq f(a')$. Exercise 1.2.22 asks you to verify this.

Sometimes we meet functor-like operations that reverse the arrows, with a map $A \to A'$ in $\mathscr{A}$ giving rise to a map $F(A) \leftarrow F(A')$ in $\mathscr{B}$. Such operations are called contravariant functors.

Definition 1.2.10   Let $\mathscr{A}$ and $\mathscr{B}$ be categories. A **contravariant functor** from $\mathscr{A}$ to $\mathscr{B}$ is a functor $\mathscr{A}^{\mathrm{op}} \to \mathscr{B}$.

To avoid confusion, we write 'a contravariant functor from $\mathscr{A}$ to $\mathscr{B}$' rather than 'a contravariant functor $\mathscr{A} \to \mathscr{B}$'.

Functors $\mathscr{C} \to \mathscr{D}$ correspond one-to-one with functors $\mathscr{C}^{\mathrm{op}} \to \mathscr{D}^{\mathrm{op}}$, and $(\mathscr{A}^{\mathrm{op}})^{\mathrm{op}} = \mathscr{A}$, so a contravariant functor from $\mathscr{A}$ to $\mathscr{B}$ can also be described as a functor $\mathscr{A} \to \mathscr{B}^{\mathrm{op}}$. Which description we use is not enormously important, but in the long run, the convention in Definition 1.2.10 makes life easier.

An ordinary functor $\mathscr{A} \to \mathscr{B}$ is sometimes called a **covariant functor** from $\mathscr{A}$ to $\mathscr{B}$, for emphasis.

Example 1.2.11   We can tell a lot about a space by examining the functions on it. The importance of this principle in twentieth- and twenty-first-century mathematics can hardly be exaggerated.

For example, given a topological space X, let $C(X)$ be the ring of continuous real-valued functions on X. The ring operations are defined 'pointwise': for instance, if $p_1, p_2 \colon X \to \mathbb{R}$ are continuous maps then the map $p_1 + p_2 \colon X \to \mathbb{R}$ is defined by

$$(p_1 + p_2)(x) = p_1(x) + p_2(x)$$

($x \in X$). A continuous map $f \colon X \to Y$ induces a ring homomorphism $C(f) \colon C(Y) \to C(X)$, defined at $q \in C(Y)$ by taking $(C(f))(q)$ to be the composite map

$$X \xrightarrow{f} Y \xrightarrow{q} \mathbb{R}.$$

Note that $C(f)$ goes in the opposite direction from f. After checking some axioms (Exercise 1.2.26), we conclude that C is a contravariant functor from **Top** to **Ring**.

While this particular example will not play a large part in this text, it is worth close attention. It illustrates the important idea of a structure whose elements are maps (in this case, a ring whose elements are continuous functions). The way in which C becomes a functor, via composition, is also important. Similar constructions will be crucial in later chapters.

For certain classes of space, the passage from X to $C(X)$ loses no information: there is a way of reconstructing the space X from the ring $C(X)$. For this and related reasons, it is sometimes said that 'algebra is dual to geometry'.

Example 1.2.12 Let k be a field. For any two vector spaces V and W over k, there is a vector space

$$\mathbf{Hom}(V, W) = \{\text{linear maps } V \to W\}.$$

The elements of this vector space are themselves maps, and the vector space operations (addition and scalar multiplication) are defined pointwise, as in the last example.

Now fix a vector space W. Any linear map $f \colon V \to V'$ induces a linear map

$$f^* \colon \mathbf{Hom}(V', W) \to \mathbf{Hom}(V, W),$$

defined at $q \in \mathbf{Hom}(V', W)$ by taking $f^*(q)$ to be the composite map

$$V \xrightarrow{f} V' \xrightarrow{q} W.$$

This defines a functor

$$\mathbf{Hom}(-, W) \colon \mathbf{Vect}_k^{\mathrm{op}} \to \mathbf{Vect}_k.$$

The symbol '$-$' is a blank or placeholder, into which arguments can be inserted. Thus, the value of $\mathbf{Hom}(-, W)$ at V is $\mathbf{Hom}(V, W)$. Sometimes we use a blank space instead of $-$, as in $\mathbf{Hom}(\quad, W)$.

An important special case is where W is k, seen as a one-dimensional vector space over itself. The vector space $\mathbf{Hom}(V, k)$ is called the **dual** of V, and is written as V^*. So there is a contravariant functor

$$(\ )^* = \mathbf{Hom}(-, k) \colon \mathbf{Vect}_k^{\mathrm{op}} \to \mathbf{Vect}_k$$

sending each vector space to its dual.

Example 1.2.13　For each $n \in \mathbb{N}$, there is a functor $H^n \colon \mathbf{Top}^{\mathrm{op}} \to \mathbf{Ab}$ assigning to a space its nth cohomology group.

Example 1.2.14　Let G be a monoid, regarded as a one-object category $\mathscr{G}$. A functor $\mathscr{G}^{\mathrm{op}} \to \mathbf{Set}$ is a *right* G-set, for essentially the same reasons as in Example 1.2.8.

That left actions are covariant functors and right actions are contravariant functors is a consequence of a basic notational choice: we write the value of a function f at an element x as $f(x)$, not $(x)f$.

Contravariant functors whose codomain is **Set** are important enough to have their own special name.

Definition 1.2.15　Let $\mathscr{A}$ be a category. A **presheaf** on $\mathscr{A}$ is a functor $\mathscr{A}^{\mathrm{op}} \to$ **Set**.

The name comes from the following special case. Let X be a topological space. Write $\mathscr{O}(X)$ for the poset of open subsets of X, ordered by inclusion. View $\mathscr{O}(X)$ as a category, as in Example 1.1.8(e). Thus, the objects of $\mathscr{O}(X)$ are the open subsets of X, and for $U, U' \in \mathscr{O}(X)$, there is one map $U \to U'$ if $U \subseteq U'$, and there are none otherwise. A **presheaf** on the space X is a presheaf on the category $\mathscr{O}(X)$. For example, given any space X, there is a presheaf F on X defined by

$$F(U) = \{\text{continuous functions } U \to \mathbb{R}\}$$

($U \in \mathscr{O}(X)$) and, whenever $U \subseteq U'$ are open subsets of X, by taking the map $F(U') \to F(U)$ to be restriction. Presheaves, and a certain class of presheaves called sheaves, play an important role in modern geometry.

We know very well that for *functions* between *sets*, it is sometimes useful to consider special kinds of function such as injections, surjections and bijections. We also know that the notions of injection and subset are related: for instance,

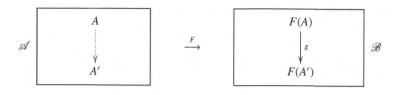

Figure 1.1 Fullness and faithfulness.

whenever B is a subset of A, there is an injection $B \to A$ given by inclusion. In this section and the next, we introduce some similar notions for *functors* between *categories*, beginning with the following definitions.

Definition 1.2.16 A functor $F \colon \mathscr{A} \to \mathscr{B}$ is **faithful** (respectively, **full**) if for each $A, A' \in \mathscr{A}$, the function

$$\mathscr{A}(A, A') \quad \to \quad \mathscr{B}(F(A), F(A'))$$
$$f \quad \mapsto \quad F(f)$$

is injective (respectively, surjective).

Warning 1.2.17 Note the roles of A and A' in the definition. Faithfulness does *not* say that if f_1 and f_2 are distinct maps in $\mathscr{A}$ then $F(f_1) \neq F(f_2)$ (Exercise 1.2.27). In the situation of Figure 1.1, F is faithful if for each A, A' and g as shown, there is at most one dotted arrow that F sends to g. It is full if for each such A, A' and g, there is at least one dotted arrow that F sends to g.

Definition 1.2.18 Let $\mathscr{A}$ be a category. A **subcategory** $\mathscr{S}$ of $\mathscr{A}$ consists of a subclass $\mathrm{ob}(\mathscr{S})$ of $\mathrm{ob}(\mathscr{A})$ together with, for each $S, S' \in \mathrm{ob}(\mathscr{S})$, a subclass $\mathscr{S}(S, S')$ of $\mathscr{A}(S, S')$, such that $\mathscr{S}$ is closed under composition and identities. It is a **full** subcategory if $\mathscr{S}(S, S') = \mathscr{A}(S, S')$ for all $S, S' \in \mathrm{ob}(\mathscr{S})$.

A full subcategory therefore consists of a selection of the objects, with all of the maps between them. So, a full subcategory can be specified simply by saying what its objects are. For example, **Ab** is the full subcategory of **Grp** consisting of the groups that are abelian.

Whenever $\mathscr{S}$ is a subcategory of a category $\mathscr{A}$, there is an inclusion functor $I \colon \mathscr{S} \to \mathscr{A}$ defined by $I(S) = S$ and $I(f) = f$. It is automatically faithful, and it is full if and only if $\mathscr{S}$ is a full subcategory.

Warning 1.2.19 The image of a functor need not be a subcategory. For example, consider the functor

$$\left(A \xrightarrow{\ f\ } B \quad B' \xrightarrow{\ g\ } C \right) \quad \xrightarrow{\ F\ } \quad \left(\begin{array}{c} Y \\ p \nearrow \quad \searrow q \\ X \xrightarrow[qp]{} Z \end{array} \right)$$

defined by $F(A) = X$, $F(B) = F(B') = Y$, $F(C) = Z$, $F(f) = p$, and $F(g) = q$. Then p and q are in the image of F, but qp is not.

Exercises

1.2.20 Find three examples of functors not mentioned above.

1.2.21 Show that functors preserve isomorphism. That is, prove that if $F: \mathscr{A} \to \mathscr{B}$ is a functor and $A, A' \in \mathscr{A}$ with $A \cong A'$, then $F(A) \cong F(A')$.

1.2.22 Prove the assertion made in Example 1.2.9. In other words, given ordered sets A and B, and denoting by $\mathscr{A}$ and $\mathscr{B}$ the corresponding categories, show that a functor $\mathscr{A} \to \mathscr{B}$ amounts to an order-preserving map $A \to B$.

1.2.23 Two categories $\mathscr{A}$ and $\mathscr{B}$ are **isomorphic**, written as $\mathscr{A} \cong \mathscr{B}$, if they are isomorphic as objects of **CAT**.

(a) Let G be a group, regarded as a one-object category all of whose maps are isomorphisms. Then its opposite G^{op} is also a one-object category all of whose maps are isomorphisms, and can therefore be regarded as a group too. What is G^{op}, in purely group-theoretic terms? Prove that G is isomorphic to G^{op}.

(b) Find a monoid not isomorphic to its opposite.

1.2.24 Is there a functor $Z: \mathbf{Grp} \to \mathbf{Grp}$ with the property that $Z(G)$ is the centre of G for all groups G?

1.2.25 Sometimes we meet functors whose domain is a product $\mathscr{A} \times \mathscr{B}$ of categories. Here you will show that such a functor can be regarded as an interlocking pair of families of functors, one defined on $\mathscr{A}$ and the other defined on $\mathscr{B}$. (This is very like the situation for bilinear and linear maps.)

(a) Let $F: \mathscr{A} \times \mathscr{B} \to \mathscr{C}$ be a functor. Prove that for each $A \in \mathscr{A}$, there is a functor $F^A: \mathscr{B} \to \mathscr{C}$ defined on objects $B \in \mathscr{B}$ by $F^A(B) = F(A, B)$ and on maps g in $\mathscr{B}$ by $F^A(g) = F(1_A, g)$. Prove that for each $B \in \mathscr{B}$, there is a functor $F_B: \mathscr{A} \to \mathscr{C}$ defined similarly.

(b) Let $F: \mathscr{A} \times \mathscr{B} \to \mathscr{C}$ be a functor. With notation as in (a), show that the families of functors $(F^A)_{A \in \mathscr{A}}$ and $(F_B)_{B \in \mathscr{B}}$ satisfy the following two conditions:

- if $A \in \mathscr{A}$ and $B \in \mathscr{B}$ then $F^A(B) = F_B(A)$;
- if $f: A \to A'$ in $\mathscr{A}$ and $g: B \to B'$ in $\mathscr{B}$ then $F^{A'}(g) \circ F_B(f) = F_{B'}(f) \circ F^A(g)$.

(c) Now take categories $\mathscr{A}$, $\mathscr{B}$ and $\mathscr{C}$, and take families of functors $(F^A)_{A \in \mathscr{A}}$ and $(F_B)_{B \in \mathscr{B}}$ satisfying the two conditions in (b). Prove that there is a unique functor $F: \mathscr{A} \times \mathscr{B} \to \mathscr{C}$ satisfying the equations in (a). ('There is a unique functor' means in particular that there *is* a functor, so you have to prove existence as well as uniqueness.)

1.2.26 Fill in the details of Example 1.2.11, thus constructing a functor C: **Top**$^{\mathrm{op}} \to$ **Ring**.

1.2.27 Find an example of a functor $F: \mathscr{A} \to \mathscr{B}$ such that F is faithful but there exist distinct maps f_1 and f_2 in $\mathscr{A}$ with $F(f_1) = F(f_2)$.

1.2.28 (a) Of the examples of functors appearing in this section, which are faithful and which are full?

(b) Write down one example of a functor that is both full and faithful, one that is full but not faithful, one that is faithful but not full, and one that is neither.

1.2.29 (a) What are the subcategories of an ordered set? Which are full?

(b) What are the subcategories of a group? (Careful!) Which are full?

1.3 Natural transformations

We now know about categories. We also know about functors, which are maps between categories. Perhaps surprisingly, there is a further notion of 'map between functors'. Such maps are called natural transformations. This notion only applies when the functors have the same domain and codomain:

$$\mathscr{A} \underset{G}{\overset{F}{\rightrightarrows}} \mathscr{B} \;.$$

To see how this might work, let us consider a special case. Let $\mathscr{A}$ be the discrete category (Example 1.1.8(b)) whose objects are the natural numbers $0, 1, 2, \ldots$. A functor F from $\mathscr{A}$ to another category $\mathscr{B}$ is simply a sequence

$(F_0, F_1, F_2, \ldots)$ of objects of $\mathscr{B}$. Let G be another functor from $\mathscr{A}$ to $\mathscr{B}$, consisting of another sequence $(G_0, G_1, G_2, \ldots)$ of objects of $\mathscr{B}$. It would be reasonable to define a 'map' from F to G to be a sequence

$$\left(F_0 \xrightarrow{\alpha_0} G_0, \ F_1 \xrightarrow{\alpha_1} G_1, \ F_1 \xrightarrow{\alpha_2} G_2, \ \ldots \right)$$

of maps in $\mathscr{B}$. The situation can be depicted as follows:

$$\mathscr{A} \quad \boxed{\begin{array}{cccc} 0 & 1 & 2 & \cdots \end{array}} \qquad \boxed{\begin{array}{cccc} F_0 & F_1 & F_2 & \\ \downarrow{\scriptstyle\alpha_0} & \downarrow{\scriptstyle\alpha_1} & \downarrow{\scriptstyle\alpha_2} & \cdots \\ G_0 & G_1 & G_2 & \end{array}} \quad \mathscr{B}$$

(The right-hand diagram should not be understood too literally. Some of the objects F_i or G_i might be equal, and there might be much else in $\mathscr{B}$ besides what is shown.)

This suggests that in the general case, a natural transformation between functors $\mathscr{A} \overset{F}{\underset{G}{\rightrightarrows}} \mathscr{B}$ should consist of maps $\alpha_A\colon F(A) \to G(A)$, one for each $A \in \mathscr{A}$. In the example above, the category $\mathscr{A}$ had the special property of not containing any nontrivial maps. In general, we demand some kind of compatibility between the maps in $\mathscr{A}$ and the maps α_A.

Definition 1.3.1   Let $\mathscr{A}$ and $\mathscr{B}$ be categories and let $\mathscr{A} \overset{F}{\underset{G}{\rightrightarrows}} \mathscr{B}$ be functors. A **natural transformation** $\alpha\colon F \to G$ is a family $\left(F(A) \xrightarrow{\alpha_A} G(A)\right)_{A \in \mathscr{A}}$ of maps in $\mathscr{B}$ such that for every map $A \xrightarrow{f} A'$ in $\mathscr{A}$, the square

$$\begin{array}{ccc} F(A) & \xrightarrow{\ F(f)\ } & F(A') \\ {\scriptstyle\alpha_A}\downarrow & & \downarrow{\scriptstyle\alpha_{A'}} \\ G(A) & \xrightarrow[\ G(f)\ ]{} & G(A') \end{array} \qquad (1.3)$$

commutes. The maps α_A are called the **components** of α.

Remarks 1.3.2   (a) The definition of natural transformation is set up so that from each map $A \xrightarrow{f} A'$ in $\mathscr{A}$, it is possible to construct exactly one map $F(A) \to G(A')$ in $\mathscr{B}$. When $f = 1_A$, this map is α_A. For a general f, it is the diagonal of the square (1.3), and 'exactly one' implies that the square commutes.

(b) We write

$$\mathscr{A} \underset{G}{\overset{F}{\Longrightarrow}} \mathscr{B}$$

to mean that α is a natural transformation from F to G.

Example 1.3.3   Let $\mathscr{A}$ be a discrete category, and let $F, G\colon \mathscr{A} \to \mathscr{B}$ be functors. Then F and G are just families $(F(A))_{A\in\mathscr{A}}$ and $(G(A))_{A\in\mathscr{A}}$ of objects of $\mathscr{B}$. A natural transformation $\alpha\colon F \to G$ is just a family $\left(F(A) \overset{\alpha_A}{\longrightarrow} G(A)\right)_{A\in\mathscr{A}}$ of maps in $\mathscr{B}$, as claimed above in the case ob $\mathscr{A} = \mathbb{N}$. In principle, this family must satisfy the naturality axiom (1.3) for every map f in $\mathscr{A}$; but the only maps in $\mathscr{A}$ are the identities, and when f is an identity, this axiom holds automatically.

Example 1.3.4   Recall from Examples 1.1.8 that a group (or more generally, a monoid) G can be regarded as a one-object category. Also recall from Example 1.2.8 that a functor from the category G to **Set** is nothing but a left G-set. (Previously we used $\mathscr{G}$ to denote the category corresponding to the group G; from now on we use G to denote them both.) Take two G-sets, S and T. Since S and T can be regarded as functors $G \to$ **Set**, we can ask: what is a natural transformation

$$G \underset{T}{\overset{S}{\Longrightarrow}} \textbf{Set},$$

in concrete terms?

Such a natural transformation consists of a single map in **Set** (since G has just one object), satisfying some axioms. Precisely, it is a function $\alpha\colon S \to T$ such that $\alpha(g \cdot s) = g \cdot \alpha(s)$ for all $s \in S$ and $g \in G$. (Why?) In other words, it is just a map of G-sets, sometimes called a G-**equivariant** map.

Example 1.3.5   Fix a natural number n. In this example, we will see how 'determinant of an $n \times n$ matrix' can be understood as a natural transformation.

For any commutative ring R, the $n \times n$ matrices with entries in R form a monoid $M_n(R)$ under multiplication. Moreover, any ring homomorphism $R \to S$ induces a monoid homomorphism $M_n(R) \to M_n(S)$. This defines a functor $M_n\colon$ **CRing** $\to$ **Mon** from the category of commutative rings to the category of monoids.

Also, the elements of any ring R form a monoid $U(R)$ under multiplication, giving another functor $U\colon$ **CRing** $\to$ **Mon**.

Now, every $n \times n$ matrix X over a commutative ring R has a determinant $\det_R(X)$, which is an element of R. Familiar properties of determinant –

$$\det_R(XY) = \det_R(X)\det_R(Y), \qquad \det_R(I) = 1$$

– tell us that for each R, the function $\det_R \colon M_n(R) \to U(R)$ is a monoid homomorphism. So, we have a family of maps

$$\left(M_n(R) \xrightarrow{\det_R} U(R) \right)_{R \in \mathbf{CRing}},$$

and it makes sense to ask whether they define a natural transformation

$$\mathbf{CRing} \overset{M_n}{\underset{U}{\Longrightarrow}} \Downarrow \det \; \mathbf{Mon}.$$

Indeed, they do. That the naturality squares commute (check!) reflects the fact that determinant is defined in the same way for all rings. We do not use one definition of determinant for one ring and a different definition for another ring. Generally speaking, the naturality axiom (1.3) is supposed to capture the idea that the family $(\alpha_A)_{A \in \mathscr{A}}$ is defined in a uniform way across all $A \in \mathscr{A}$.

Construction 1.3.6   Natural transformations are a kind of map, so we would expect to be able to compose them. We can. Given natural transformations

$$\mathscr{A} \;\substack{\xrightarrow{F} \\ \Downarrow \alpha \\ \xrightarrow{G} \\ \Downarrow \beta \\ \xrightarrow{H}}\; \mathscr{B} \,,$$

there is a composite natural transformation

$$\mathscr{A} \;\substack{\xrightarrow{F} \\ \Downarrow \beta \circ \alpha \\ \xrightarrow{H}}\; \mathscr{B}$$

defined by $(\beta \circ \alpha)_A = \beta_A \circ \alpha_A$ for all $A \in \mathscr{A}$. There is also an identity natural transformation

$$\mathscr{A} \;\substack{\xrightarrow{F} \\ \Downarrow 1_F \\ \xrightarrow{F}}\; \mathscr{B}$$

on any functor F, defined by $(1_F)_A = 1_{F(A)}$. So for any two categories $\mathscr{A}$ and $\mathscr{B}$, there is a category whose objects are the functors from $\mathscr{A}$ to $\mathscr{B}$ and whose maps are the natural transformations between them. This is called the **functor category** from $\mathscr{A}$ to $\mathscr{B}$, and written as $[\mathscr{A}, \mathscr{B}]$ or $\mathscr{B}^{\mathscr{A}}$.

Example 1.3.7 Let 2 be the discrete category with two objects. A functor from 2 to a category $\mathscr{B}$ is a pair of objects of $\mathscr{B}$, and a natural transformation is a pair of maps. The functor category $[2, \mathscr{B}]$ is therefore isomorphic to the product category $\mathscr{B} \times \mathscr{B}$ (Construction 1.1.11). This fits well with the alternative notation $\mathscr{B}^2$ for the functor category.

Example 1.3.8 Let G be a monoid. Then $[G, \mathbf{Set}]$ is the category of left G-sets, and $[G^{\mathrm{op}}, \mathbf{Set}]$ is the category of right G-sets (Example 1.2.14).

Example 1.3.9 Take ordered sets A and B, viewed as categories (as in Example 1.1.8(e)). Given order-preserving maps $A \underset{g}{\overset{f}{\rightrightarrows}} B$, viewed as functors (as in Example 1.2.9), there is at most one natural transformation

$$A \underset{g}{\overset{f}{\Downarrow}} B,$$

and there is one if and only if $f(a) \leq g(a)$ for all $a \in A$. (The naturality axiom (1.3) holds automatically, because in an ordered set, all diagrams commute.) So $[A, B]$ is an ordered set too; its elements are the order-preserving maps from A to B, and $f \leq g$ if and only if $f(a) \leq g(a)$ for all $a \in A$.

Everyday phrases such as '*the* cyclic group of order 6' and '*the* product of two spaces' reflect the fact that given two isomorphic objects of a category, we usually neither know nor care whether they are actually equal. This is enormously important.

In particular, the lesson applies when the category concerned is a functor category. In other words, given two functors $F, G: \mathscr{A} \to \mathscr{B}$, we usually do not care whether they are literally equal. (Equality would imply that the objects $F(A)$ and $G(A)$ of $\mathscr{B}$ were equal for all $A \in \mathscr{A}$, a level of detail in which we have just declared ourselves to be uninterested.) What really matters is whether they are naturally isomorphic.

Definition 1.3.10 Let $\mathscr{A}$ and $\mathscr{B}$ be categories. A **natural isomorphism** between functors from $\mathscr{A}$ to $\mathscr{B}$ is an isomorphism in $[\mathscr{A}, \mathscr{B}]$.

An equivalent form of the definition is often useful:

Lemma 1.3.11 *Let* $\mathscr{A} \underset{G}{\overset{F}{\Downarrow \alpha}} \mathscr{B}$ *be a natural transformation. Then α is a natural isomorphism if and only if $\alpha_A \colon F(A) \to G(A)$ is an isomorphism for all $A \in \mathscr{A}$.*

Proof   Exercise 1.3.26.   □

Of course, we say that functors F and G are **naturally isomorphic** if there exists a natural isomorphism from F to G. Since natural isomorphism is just isomorphism in a particular category (namely, $[\mathscr{A}, \mathscr{B}]$), we already have notation for this: $F \cong G$.

Definition 1.3.12   Given functors $\mathscr{A} \underset{G}{\overset{F}{\rightrightarrows}} \mathscr{B}$, we say that

$$F(A) \cong G(A) \text{ **naturally in** } A$$

if F and G are naturally isomorphic.

This alternative terminology can be understood as follows. If $F(A) \cong G(A)$ naturally in A then certainly $F(A) \cong G(A)$ for each individual A, but more is true: we can choose isomorphisms $\alpha_A \colon F(A) \to G(A)$ in such a way that the naturality axiom (1.3) is satisfied.

Example 1.3.13   Let $F, G \colon \mathscr{A} \to \mathscr{B}$ be functors from a discrete category $\mathscr{A}$ to a category $\mathscr{B}$. Then $F \cong G$ if and only if $F(A) \cong G(A)$ for all $A \in \mathscr{A}$.

So in *this* case, $F(A) \cong G(A)$ naturally in A if and only if $F(A) \cong G(A)$ for all A. But this is only true because $\mathscr{A}$ is discrete. In general, it is emphatically false. There are many examples of categories and functors $\mathscr{A} \underset{G}{\overset{F}{\rightrightarrows}} \mathscr{B}$ such that $F(A) \cong G(A)$ for all $A \in \mathscr{A}$, but not *naturally* in A. Exercise 1.3.31 gives an example from combinatorics.

Example 1.3.14   Let **FDVect** be the category of finite-dimensional vector spaces over some field k. The dual vector space construction defines a contravariant functor from **FDVect** to itself (Example 1.2.12), and the double dual construction therefore defines a covariant functor from **FDVect** to itself.

Moreover, we have for each $V \in$ **FDVect** a canonical isomorphism $\alpha_V \colon V \to V^{**}$. Given $v \in V$, the element $\alpha_V(v)$ of V^{**} is 'evaluation at v'; that is, $\alpha_V(v) \colon V^* \to k$ maps $\phi \in V^*$ to $\phi(v) \in k$. That α_V is an isomorphism is a standard result in the theory of finite-dimensional vector spaces.

This defines a natural transformation

$$\textbf{FDVect} \underset{(\ )^{**}}{\overset{1_{\textbf{FDVect}}}{\Downarrow \alpha}} \textbf{FDVect}$$

from the identity functor to the double dual functor. By Lemma 1.3.11, α is a natural isomorphism. So $1_{\mathbf{FDVect}} \cong (\ )^{**}$. Equivalently, in the language of Definition 1.3.12, $V \cong V^{**}$ naturally in V.

This is one of those occasions on which category theory makes an intuition precise. In some informal sense, evident before you learn anything about category theory, the isomorphism between a finite-dimensional vector space and its double dual is 'natural' or 'canonical': no arbitrary choices are needed in order to define it. In contrast, to specify an isomorphism between V and its single dual V^*, we need to make an arbitrary choice of basis, and the isomorphism really does depend on the basis that we choose.

In the example on vector spaces, the word **canonical** was used. It is an informal word, meaning something like 'God-given' or 'defined without making arbitrary choices'. For example, for any two sets A and B, there is a canonical bijection $A \times B \to B \times A$ defined by $(a, b) \mapsto (b, a)$, and there is a canonical function $A \times B \to A$ defined by $(a, b) \mapsto a$. But the function $B \to A$ defined by 'choose an element $a_0 \in A$ and send everything to a_0' is not canonical, because the choice of a_0 is arbitrary.

The concept of natural isomorphism leads unavoidably to another central concept: equivalence of categories.

Two elements of a set are either equal or not. Two objects of a category can be equal, not equal but isomorphic, or not even isomorphic. As explained before Definition 1.3.10, the notion of equality between two objects of a category is unreasonably strict; it is usually isomorphism that we care about. So:

- the right notion of sameness of two elements of a set is equality;
- the right notion of sameness of two objects of a category is isomorphism.

When applied to a functor category $[\mathscr{A}, \mathscr{B}]$, the second point tells us that:

- the right notion of sameness of two functors $\mathscr{A} \rightrightarrows \mathscr{B}$ is natural isomorphism.

But what is the right notion of sameness of two *categories*? Isomorphism is unreasonably strict, as if $\mathscr{A} \cong \mathscr{B}$ then there are functors

$$\mathscr{A} \underset{G}{\overset{F}{\rightleftarrows}} \mathscr{B} \tag{1.4}$$

such that

$$G \circ F = 1_{\mathscr{A}} \quad \text{and} \quad F \circ G = 1_{\mathscr{B}}, \tag{1.5}$$

and we have just seen that the notion of equality between functors is too strict. The most useful notion of sameness of categories, called 'equivalence', is looser than isomorphism. To obtain the definition, we simply replace the unreasonably strict equalities in (1.5) by isomorphisms. This gives

$$G \circ F \cong 1_{\mathscr{A}} \quad \text{and} \quad F \circ G \cong 1_{\mathscr{B}}.$$

Definition 1.3.15 An **equivalence** between categories $\mathscr{A}$ and $\mathscr{B}$ consists of a pair (1.4) of functors together with natural isomorphisms

$$\eta: 1_{\mathscr{A}} \to G \circ F, \quad \varepsilon: F \circ G \to 1_{\mathscr{B}}.$$

If there exists an equivalence between $\mathscr{A}$ and $\mathscr{B}$, we say that $\mathscr{A}$ and $\mathscr{B}$ are **equivalent**, and write $\mathscr{A} \simeq \mathscr{B}$. We also say that the functors F and G are **equivalences**.

The directions of η and ε are not very important, since they are isomorphisms anyway. The reason for this particular choice will become apparent when we come to discuss adjunctions (Section 2.2).

Warning 1.3.16 The symbol $\cong$ is used for isomorphism of objects of a category, and in particular for isomorphism of categories (which are objects of **CAT**). The symbol $\simeq$ is used for equivalence of categories. At least, this is the convention used in this book and by most category theorists, although it is far from universal in mathematics at large.

There is a very useful alternative characterization of those functors that are equivalences. First, we need a definition.

Definition 1.3.17 A functor $F: \mathscr{A} \to \mathscr{B}$ is **essentially surjective on objects** if for all $B \in \mathscr{B}$, there exists $A \in \mathscr{A}$ such that $F(A) \cong B$.

Proposition 1.3.18 *A functor is an equivalence if and only if it is full, faithful and essentially surjective on objects.*

Proof Exercise 1.3.32. □

This result can be compared to the theorem that every bijective group homomorphism is an isomorphism (that is, its inverse is also a homomorphism), or that a natural transformation whose components are isomorphisms is itself an isomorphism (Lemma 1.3.11). Those two results are useful because they allow us to show that a map is an isomorphism without directly constructing an inverse. Proposition 1.3.18 provides a similar service, enabling us to prove that a functor F is an equivalence without actually constructing an 'inverse' G, or indeed an η or an ε (in the notation of Definition 1.3.15).

A corollary of Proposition 1.3.18 invites us to view full and faithful functors as, essentially, inclusions of full subcategories:

Corollary 1.3.19 *Let* $F: \mathscr{C} \to \mathscr{D}$ *be a full and faithful functor. Then* $\mathscr{C}$ *is equivalent to the full subcategory* $\mathscr{C}'$ *of* $\mathscr{D}$ *whose objects are those of the form* $F(C)$ *for some* $C \in \mathscr{C}$.

Proof The functor $F': \mathscr{C} \to \mathscr{C}'$ defined by $F'(C) = F(C)$ is full and faithful (since F is) and essentially surjective on objects (by definition of $\mathscr{C}'$). □

This result is true, with the same proof, whether we interpret 'of the form $F(C)$' to mean 'equal to $F(C)$' or 'isomorphic to $F(C)$'.

Example 1.3.20 Let $\mathscr{A}$ be any category, and let $\mathscr{B}$ be any full subcategory containing at least one object from each isomorphism class of $\mathscr{A}$. Then the inclusion functor $\mathscr{B} \hookrightarrow \mathscr{A}$ is faithful (like any inclusion of subcategories), full, and essentially surjective on objects. Hence $\mathscr{B} \simeq \mathscr{A}$.

So if we take a category and remove some (but not all) of the objects in each isomorphism class, the slimmed-down version is equivalent to the original. Conversely, if we take a category and throw in some more objects, each of them isomorphic to one of the existing objects, it makes no difference: the new, bigger, category is equivalent to the old one.

For example, let **FinSet** be the category of finite sets and functions between them. For each natural number n, choose a set $\mathbf{n}$ with n elements, and let $\mathscr{B}$ be the full subcategory of **FinSet** with objects $\mathbf{0}, \mathbf{1}, \dots$. Then $\mathscr{B} \simeq$ **FinSet**, even though $\mathscr{B}$ is in some sense much smaller than **FinSet**.

Example 1.3.21 In Example 1.1.8(d), we saw that monoids are essentially the same thing as one-object categories. With the definition of equivalence in hand, we are nearly ready to make this statement precise. We are missing some set-theoretic language, and we will return to this result once we have that language (Example 3.2.11), but the essential point can be stated now.

Let $\mathscr{C}$ be the full subcategory of **CAT** whose objects are the one-object categories. Let **Mon** be the category of monoids. Then $\mathscr{C} \simeq$ **Mon**. To see this, first note that given any object A of any category, the maps $A \to A$ form a monoid under composition (at least, subject to some set-theoretic restrictions). There is, therefore, a canonical functor $F : \mathscr{C} \to$ **Mon** sending a one-object category to the monoid of maps from the single object to itself. This functor F is full and faithful (by Example 1.2.7) and essentially surjective on objects. Hence F is an equivalence.

Example 1.3.22 An equivalence of the form $\mathscr{A}^{\mathrm{op}} \simeq \mathscr{B}$ is sometimes called a **duality** between $\mathscr{A}$ and $\mathscr{B}$. One says that $\mathscr{A}$ is **dual** to $\mathscr{B}$. There are many famous dualities in which $\mathscr{A}$ is a category of algebras and $\mathscr{B}$ is a category of spaces; recall the slogan 'algebra is dual to geometry' from Example 1.2.11.

Here are some quite advanced examples, well beyond the scope of this book.

- Stone duality: the category of Boolean algebras is dual to the category of totally disconnected compact Hausdorff spaces.
- Gelfand–Naimark duality: the category of commutative unital C^*-algebras is dual to the category of compact Hausdorff spaces. (C^*-algebras are certain algebraic structures important in functional analysis.)
- Algebraic geometers have several notions of 'space', one of which is 'affine variety'. Let k be an algebraically closed field. Then the category of affine varieties over k is dual to the category of finitely generated k-algebras with no nontrivial nilpotents.
- Pontryagin duality: the category of locally compact abelian topological groups is dual to itself. As the words 'topological group' suggest, both sides of the duality are algebraic *and* geometric. Pontryagin duality is an abstraction of the properties of the Fourier transform.

Example 1.3.23 It is rarely useful to consider a category of structured objects in which the maps do not respect that structure. For instance, let $\mathscr{A}$ be the category whose objects are groups and whose maps are *all* functions between them, not necessarily homomorphisms. Let $\mathbf{Set}_{\neq\emptyset}$ be the category of nonempty sets. The forgetful functor $U \colon \mathscr{A} \to \mathbf{Set}_{\neq\emptyset}$ is full and faithful. It is a (not profound) fact that every nonempty set can be given at least one group structure, so U is essentially surjective on objects. Hence U is an equivalence. This implies that the category $\mathscr{A}$, although defined in terms of groups, is really just the category of nonempty sets.

Remarks 1.3.24 Here is a kind of review of the chapter so far. We have defined:

- categories (Section 1.1);
- functors between categories (Section 1.2);
- natural transformations between functors (Section 1.3);
- composition of functors

$$\cdot \to \cdot \to \cdot$$

and the identity functor on any category (Remark 1.2.2(b));
- composition of natural transformations

and the identity natural transformation on any functor (Construction 1.3.6).

This composition of natural transformations is sometimes called **vertical composition**. There is also **horizontal composition**, which takes natural transformations

$$\mathscr{A} \xRightarrow[G]{F} \alpha\ \mathscr{A}' \xRightarrow[G']{F'} \alpha'\ \mathscr{A}''$$

and produces a natural transformation

$$\mathscr{A} \xRightarrow[G'\circ G]{F'\circ F} \mathscr{A}'',$$

traditionally written as $\alpha' * \alpha$. The component of $\alpha' * \alpha$ at $A \in \mathscr{A}$ is defined to be the diagonal of the naturality square

$$
\begin{array}{ccc}
F'(F(A)) & \xrightarrow{F'(\alpha_A)} & F'(G(A)) \\
{\scriptstyle \alpha'_{F(A)}}\downarrow & & \downarrow{\scriptstyle \alpha'_{G(A)}} \\
G'(F(A)) & \xrightarrow[G'(\alpha_A)]{} & G'(G(A)).
\end{array}
$$

In other words, $(\alpha' * \alpha)_A$ can be defined as either $\alpha'_{G(A)} \circ F'(\alpha_A)$ or $G'(\alpha_A) \circ \alpha'_{F(A)}$; it makes no difference which, since they are equal.

The special cases of horizontal composition where either α or α' is an identity are especially important, and have their own notation. Thus,

$$\mathscr{A} \xrightarrow{F} \mathscr{A}' \xRightarrow[G']{F'} \alpha'\ \mathscr{A}'' \qquad \text{gives rise to} \qquad \mathscr{A} \xRightarrow[G'\circ F]{F'\circ F} \alpha'F\ \mathscr{A}''$$

where $(\alpha'F)_A = \alpha'_{F(A)}$, and

$$\mathscr{A} \xRightarrow[G]{F} \alpha\ \mathscr{A}' \xrightarrow{F'} \mathscr{A}'' \qquad \text{gives rise to} \qquad \mathscr{A} \xRightarrow[F'\circ G]{F'\circ F} F'\alpha\ \mathscr{A}''$$

where $(F'\alpha)_A = F'(\alpha_A)$.

Vertical and horizontal composition interact well: natural transformations

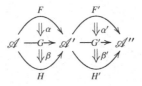

obey the **interchange law**,

$$(\beta' \circ \alpha') * (\beta \circ \alpha) = (\beta' * \beta) \circ (\alpha' * \alpha) \colon F' \circ F \to H' \circ H.$$

As usual, a statement on composition is accompanied by a statement on identities: $1_{F'} * 1_F = 1_{F' \circ F}$ too.

All of this enables us to construct, for any categories $\mathscr{A}$, $\mathscr{A}'$ and $\mathscr{A}''$, a functor

$$[\mathscr{A}', \mathscr{A}''] \times [\mathscr{A}, \mathscr{A}'] \to [\mathscr{A}, \mathscr{A}''],$$

given on objects by $(F', F) \mapsto F' \circ F$ and on maps by $(\alpha', \alpha) \mapsto \alpha' * \alpha$. In particular, if $F' \cong G'$ and $F \cong G$ then $F' \circ F \cong G' \circ G$, since functors preserve isomorphism (Exercise 1.2.21).

(The existence of this functor is similar to the fact that *inside* a category $\mathscr{C}$, we have, for any objects A, A' and A'', a func*tion*

$$\mathscr{C}(A', A'') \times \mathscr{C}(A, A') \to \mathscr{C}(A, A''),$$

given by $(f', f) \mapsto f' \circ f$.)

The diagrams above contain not only objects (0-dimensional) and arrows $\to$ (1-dimensional), but also double arrows $\Rightarrow$ sweeping out 2-dimensional regions between arrows. What we are implicitly doing is called 2-category theory. There is a 2-category of categories, functors and natural transformations, whose anatomy we have just been describing. If we are really serious about categories, we have to get serious about 2-categories. And if we are really serious about 2-categories, we have to get serious about 3-categories... and before we know it, we are studying ∞-categories. But in this book, we climb no higher than the first rung or two of this infinite ladder.

Exercises

1.3.25 Find three examples of natural transformations not mentioned above.

1.3.26 Prove Lemma 1.3.11.

1.3.27 Let $\mathscr{A}$ and $\mathscr{B}$ be categories. Prove that $[\mathscr{A}^{\mathrm{op}}, \mathscr{B}^{\mathrm{op}}] \cong [\mathscr{A}, \mathscr{B}]^{\mathrm{op}}$.

1.3.28 Let A and B be sets, and denote by B^A the set of functions from A to B. Write down:

(a) a canonical function $A \times B^A \to B$;
(b) a canonical function $A \to B^{(B^A)}$.

(Although in principle there could be many such canonical functions, in both these cases there is only one.)

1.3.29 Here we consider natural transformations between functors whose domain is a product category $\mathscr{A} \times \mathscr{B}$. Your task is to show that naturality in two variables simultaneously is equivalent to naturality in each variable separately.

Take functors $F, G: \mathscr{A} \times \mathscr{B} \to \mathscr{C}$. For each $A \in \mathscr{A}$, there are functors $F^A, G^A: \mathscr{B} \to \mathscr{C}$, as in Exercise 1.2.25. Similarly, for each $B \in \mathscr{B}$, there are functors $F_B, G_B: \mathscr{A} \to \mathscr{C}$.

Let $(\alpha_{A,B}: F(A, B) \to G(A, B))_{A \in \mathscr{A}, B \in \mathscr{B}}$ be a family of maps. Show that this family is a natural transformation $F \to G$ if and only if it satisfies the following two conditions:

- for each $A \in \mathscr{A}$, the family $(\alpha_{A,B}: F^A(B) \to G^A(B))_{B \in \mathscr{B}}$ is a natural transformation $F^A \to G^A$;
- for each $B \in \mathscr{B}$, the family $(\alpha_{A,B}: F_B(A) \to G_B(A))_{A \in \mathscr{A}}$ is a natural transformation $F_B \to G_B$.

1.3.30 Let G be a group. For each $g \in G$, there is a unique homomorphism $\phi: \mathbb{Z} \to G$ satisfying $\phi(1) = g$. Thus, elements of G are essentially the same thing as homomorphisms $\mathbb{Z} \to G$. When groups are regarded as one-object categories, homomorphisms $\mathbb{Z} \to G$ are in turn the same as functors $\mathbb{Z} \to G$. Natural isomorphism defines an equivalence relation on the set of functors $\mathbb{Z} \to G$, and, therefore, an equivalence relation on G itself. What is this equivalence relation, in purely group-theoretic terms?

(First have a guess. For a general group G, what equivalence relations on G can you think of?)

1.3.31 A **permutation** of a set X is a bijection $X \to X$. Write **Sym**(X) for the set of permutations of X. A **total order** on a set X is an order $\leq$ such that for all $x, y \in X$, either $x \leq y$ or $y \leq x$; so a total order on a finite set amounts to a way of placing its elements in sequence. Write **Ord**(X) for the set of total orders on X.

Let $\mathscr{B}$ denote the category of finite sets and bijections.

(a) Give a definition of **Sym** on maps in $\mathscr{B}$ in such a way that **Sym** becomes a functor $\mathscr{B} \to$ **Set**. Do the same for **Ord**. Both your definitions should be canonical (no arbitrary choices).

(b) Show that there is no natural transformation **Sym** $\to$ **Ord**. (Hint: consider identity permutations.)

(c) For an n-element set X, how many elements do the sets **Sym**(X) and **Ord**(X) have?

Conclude that **Sym**$(X) \cong$ **Ord**(X) for all $X \in \mathscr{B}$, but not *naturally* in $X \in \mathscr{B}$. (The moral is that for each finite set X, there are exactly as many permutations of X as there are total orders on X, but there is no natural way of matching them up.)

1.3.32 In this exercise, you will prove Proposition 1.3.18. Let $F: \mathscr{A} \to \mathscr{B}$ be a functor.

(a) Suppose that F is an equivalence. Prove that F is full, faithful and essentially surjective on objects. (Hint: prove faithfulness before fullness.)

(b) Now suppose instead that F is full, faithful and essentially surjective on objects. For each $B \in \mathscr{B}$, choose an object $G(B)$ of $\mathscr{A}$ and an isomorphism $\varepsilon_B: F(G(B)) \to B$. Prove that G extends to a functor in such a way that $(\varepsilon_B)_{B \in \mathscr{B}}$ is a natural isomorphism $FG \to 1_{\mathscr{B}}$. Then construct a natural isomorphism $1_{\mathscr{A}} \to GF$, thus proving that F is an equivalence.

1.3.33 This exercise makes precise the idea that linear algebra can equivalently be done with matrices or with linear maps.

Fix a field k. Let **Mat** be the category whose objects are the natural numbers and with

$$\mathbf{Mat}(m, n) = \{n \times m \text{ matrices over } k\}.$$

Prove that **Mat** is equivalent to **FDVect**, the category of finite-dimensional vector spaces over k. Does your equivalence involve a *canonical* functor from **Mat** to **FDVect**, or from **FDVect** to **Mat**?

(Part of the exercise is to work out what composition in the category **Mat** is supposed to be; there is only one sensible possibility. Proposition 1.3.18 makes the exercise easier.)

1.3.34 Show that equivalence of categories is an equivalence relation. (Not as obvious as it looks.)

2

Adjoints

The slogan of Saunders Mac Lane's book *Categories for the Working Mathematician* is:

Adjoint functors arise everywhere.

We will see the truth of this, meeting examples of adjoint functors from diverse parts of mathematics. To complement the understanding provided by examples, we will approach the theory of adjoints from three different directions, each of which carries its own intuition. Then we will prove that the three approaches are equivalent.

Understanding adjointness gives you a valuable addition to your mathematical toolkit. Most professional pure mathematicians know what categories and functors are, but far fewer know about adjoints. More should: adjoint functors are both common and easy, and knowing about adjoints helps you to spot patterns in the mathematical landscape.

2.1 Definition and examples

Consider a pair of functors in opposite directions, $F\colon \mathscr{A} \to \mathscr{B}$ and $G\colon \mathscr{B} \to \mathscr{A}$. Roughly speaking, F is said to be left adjoint to G if, whenever $A \in \mathscr{A}$ and $B \in \mathscr{B}$, maps $F(A) \to B$ are essentially the same thing as maps $A \to G(B)$.

Definition 2.1.1 Let $\mathscr{A} \overset{F}{\underset{G}{\rightleftarrows}} \mathscr{B}$ be categories and functors. We say that F is **left adjoint** to G, and G is **right adjoint** to F, and write $F \dashv G$, if

$$\mathscr{B}(F(A), B) \cong \mathscr{A}(A, G(B)) \tag{2.1}$$

naturally in $A \in \mathscr{A}$ and $B \in \mathscr{B}$. The meaning of 'naturally' is defined below. An **adjunction** between F and G is a choice of natural isomorphism (2.1).

'Naturally in $A \in \mathscr{A}$ and $B \in \mathscr{B}$' means that there is a specified bijection (2.1) for each $A \in \mathscr{A}$ and $B \in \mathscr{B}$, and that it satisfies a naturality axiom. To state it, we need some notation. Given objects $A \in \mathscr{A}$ and $B \in \mathscr{B}$, the correspondence (2.1) between maps $F(A) \to B$ and $A \to G(B)$ is denoted by a horizontal bar, in both directions:

$$\left(F(A) \xrightarrow{g} B\right) \mapsto \left(A \xrightarrow{\bar{g}} G(B)\right),$$
$$\left(F(A) \xrightarrow{\bar{f}} B\right) \leftarrow\!\shortmid \left(A \xrightarrow{f} G(B)\right).$$

So $\bar{\bar{f}} = f$ and $\bar{\bar{g}} = g$. We call $\bar{f}$ the **transpose** of f, and similarly for g. The naturality axiom has two parts:

$$\overline{\left(F(A) \xrightarrow{g} B \xrightarrow{q} B'\right)} = \left(A \xrightarrow{\bar{g}} G(B) \xrightarrow{G(q)} G(B')\right) \qquad (2.2)$$

(that is, $\overline{q \circ g} = G(q) \circ \bar{g}$) for all g and q, and

$$\overline{\left(A' \xrightarrow{p} A \xrightarrow{f} G(B)\right)} = \left(F(A') \xrightarrow{F(p)} F(A) \xrightarrow{\bar{f}} B\right) \qquad (2.3)$$

for all p and f. It makes no difference whether we put the long bar over the left or the right of these equations, since bar is self-inverse.

Remarks 2.1.2 (a) The naturality axiom might seem ad hoc, but we will see in Chapter 4 that it simply says that two particular functors are naturally isomorphic. In this section, we ignore the naturality axiom altogether, trusting that it embodies our usual intuitive idea of naturality: something defined without making any arbitrary choices.

(b) The naturality axiom implies that from each array of maps

$$A_0 \to \cdots \to A_n, \quad F(A_n) \to B_0, \quad B_0 \to \cdots \to B_m,$$

it is possible to construct exactly one map

$$A_0 \to G(B_m).$$

Compare the comments on the definitions of category, functor and natural transformation (Remarks 1.1.2(b), 1.2.2(a), and 1.3.2(a)).

(c) Not only do adjoint functors arise everywhere; better, whenever you see a pair of functors $\mathscr{A} \rightleftarrows \mathscr{B}$, there is an excellent chance that they are adjoint (one way round or the other).

For example, suppose you get talking to a mathematician who tells you that her work involves Lie algebras and associative algebras. You try to object that you don't know what either of those things is, but she carries on talking anyway, explaining that there's a way of turning any Lie algebra into an associative algebra, and also a way of turning any associative algebra into a Lie algebra. At this point, even without knowing what she's talking about, you should bet her that one process is adjoint to the other. This almost always works.

(d) A given functor G may or may not have a left adjoint, but if it does, it is unique up to isomorphism, so we may speak of '*the* left adjoint of G'. The same goes for right adjoints. We prove this later (Example 4.3.13).

You might ask 'what do we gain from knowing that two functors are adjoint?' The uniqueness is a crucial part of the answer. Let us return to the example of (c). It would take you only a few minutes to learn what Lie algebras are, what associative algebras are, and what the standard functor G is that turns an associative algebra into a Lie algebra. What about the functor F in the opposite direction? The description of F that you will find in most algebra books (under 'universal enveloping algebra') takes much longer to understand. However, you can bypass that process completely, just by knowing that F is the left adjoint of G. Since G can have only *one* left adjoint, this characterizes F completely. In a sense, it tells you all you need to know.

Examples 2.1.3 (Algebra: free ⊣ forgetful) Forgetful functors between categories of algebraic structures usually have left adjoints. For instance:

(a) Let k be a field. There is an adjunction

$$\mathbf{Vect}_k$$
$$F \uparrow \dashv \downarrow U$$
$$\mathbf{Set},$$

where U is the forgetful functor of Example 1.2.3(b) and F is the free functor of Example 1.2.4(c). Adjointness says that given a set S and a vector space V, a linear map $F(S) \to V$ is essentially the same thing as a function $S \to U(V)$.

We saw this in Example 0.4, but let us now check it in detail.

Fix a set S and a vector space V. Given a linear map $g\colon F(S) \to V$, we may define a map of sets $\bar{g}\colon S \to U(V)$ by $\bar{g}(s) = g(s)$ for all $s \in S$. This gives a function

$$\mathbf{Vect}_k(F(S), V) \quad \to \quad \mathbf{Set}(S, U(V))$$
$$g \quad \mapsto \quad \bar{g}.$$

In the other direction, given a map of sets $f\colon S \to U(V)$, we may define a linear map $\bar{f}\colon F(S) \to V$ by $\bar{f}(\sum_{s \in S} \lambda_s s) = \sum_{s \in S} \lambda_s f(s)$ for all formal linear combinations $\sum \lambda_s s \in F(S)$. This gives a function

$$\mathbf{Set}(S, U(V)) \quad \to \quad \mathbf{Vect}_k(F(S), V)$$
$$f \qquad \mapsto \qquad \bar{f}.$$

These two functions 'bar' are mutually inverse: for any linear map g: $F(S) \to V$, we have

$$\bar{g}\left(\sum_{s \in S} \lambda_s s\right) = \sum_{s \in S} \lambda_s \bar{g}(s) = \sum_{s \in S} \lambda_s g(s) = g\left(\sum_{s \in S} \lambda_s s\right)$$

for all $\sum \lambda_s s \in F(S)$, so $\bar{\bar{g}} = g$, and for any map of sets $f : S \to U(V)$, we have

$$\bar{\bar{f}}(s) = \bar{f}(s) = f(s)$$

for all $s \in S$, so $\bar{\bar{f}} = f$. We therefore have a canonical bijection between $\mathbf{Vect}_k(F(S), V)$ and $\mathbf{Set}(S, U(V))$ for each $S \in \mathbf{Set}$ and $V \in \mathbf{Vect}_k$, as required.

Here we have been careful to distinguish between the vector space V and its underlying set $U(V)$. Very often, though, in category theory as in mathematics at large, the symbol for a forgetful functor is omitted. In this example, that would mean dropping the U and leaving the reader to figure out whether each occurrence of V is intended to denote the vector space itself or its underlying set. We will soon start using such notational shortcuts ourselves.

(b) In the same way, there is an adjunction

$$\mathbf{Grp}$$
$$F \uparrow \dashv \downarrow U$$
$$\mathbf{Set}$$

where F and U are the free and forgetful functors of Examples 1.2.3(a) and 1.2.4(a).

The free group functor is tricky to construct explicitly. In Chapter 6, we will prove a result (the general adjoint functor theorem) guaranteeing that U and many functors like it all have left adjoints. To some extent, this removes the need to construct F explicitly, as observed in Remark 2.1.2(d). The point can be overstated: for a group theorist, the more descriptions of free groups that are available, the better. Explicit constructions really can be useful. But it is an important general principle that forgetful functors of this type always have left adjoints.

(c) There is an adjunction

$$\textbf{Ab}$$

$$F \dashv U$$

$$\textbf{Grp}$$

where U is the inclusion functor of Example 1.2.3(d). If G is a group then $F(G)$ is the **abelianization** G_{ab} of G. This is an abelian quotient group of G, with the property that every map from G to an abelian group factorizes uniquely through G_{ab}:

$$G \xrightarrow{\;\eta\;} G_{\text{ab}}$$
$$\searrow^{\forall \phi} \quad \vdots \exists! \bar{\phi}$$
$$\forall A.$$

Here η is the natural map from G to its quotient G_{ab}, and A is any abelian group. (We have adopted the abuse of notation advertised in example (a), omitting the symbol U at several places in this diagram.) The bijection

$$\textbf{Ab}(G_{\text{ab}}, A) \cong \textbf{Grp}(G, U(A))$$

is given in the left-to-right direction by $\psi \mapsto \psi \circ \eta$, and in the right-to-left direction by $\phi \mapsto \bar{\phi}$.

(To construct G_{ab}, let G' be the smallest normal subgroup of G containing $xyx^{-1}y^{-1}$ for all $x, y \in G$, and put $G_{\text{ab}} = G/G'$. The kernel of any homomorphism from G to an abelian group contains G', and the universal property follows.)

(d) There are adjunctions

$$\textbf{Grp}$$

$$F \dashv U \dashv R$$

$$\textbf{Mon}$$

between the categories of groups and monoids. The middle functor U is inclusion. The left adjoint F is, again, tricky to describe explicitly. Informally, $F(M)$ is obtained from M by throwing in an inverse to every element. (For example, if M is the additive monoid of natural numbers then $F(M)$ is the group of integers.) Again, the general adjoint functor theorem (Theorem 6.3.10) guarantees the existence of this adjoint.

This example is unusual in that forgetful functors do not usually have *right* adjoints. Here, given a monoid M, the group $R(M)$ is the submonoid of M consisting of all the invertible elements.

The category **Grp** is both a **reflective** and a **coreflective** subcategory of **Mon**. This means, by definition, that the inclusion functor **Grp** $\hookrightarrow$ **Mon** has both a left and a right adjoint. The previous example tells us that **Ab** is a reflective subcategory of **Grp**.

(e) Let **Field** be the category of fields, with ring homomorphisms as the maps. The forgetful functor **Field** $\to$ **Set** does *not* have a left adjoint. (For a proof, see Example 6.3.5.) The theory of fields is unlike the theories of groups, rings, and so on, because the operation $x \mapsto x^{-1}$ is not defined for *all* x (only for $x \neq 0$).

Remark 2.1.4 At several points in this book, we make contact with the idea of an **algebraic theory**. You already know several examples: the theory of groups is an algebraic theory, as are the theory of rings, the theory of vector spaces over $\mathbb{R}$, the theory of vector spaces over $\mathbb{C}$, the theory of monoids, and (rather trivially) the theory of sets. After reading the description below, you might conclude that the word 'theory' is overly grand, and that 'definition' would be more appropriate. Nevertheless, this is the established usage.

We will not need to define 'algebraic theory' formally, but it will be important to have the general idea. Let us begin by considering the theory of groups.

A group can be defined as a set X equipped with a function $\cdot : X \times X \to X$ (multiplication), another function $(\ )^{-1} : X \to X$ (inverse), and an element $e \in X$ (the identity), satisfying a familiar list of equations. More systematically, the three pieces of structure on X can be seen as maps of sets

$$\cdot : X^2 \to X, \qquad (\ )^{-1} : X^1 \to X, \qquad e : X^0 \to X,$$

where in the last case, X^0 is the one-element set 1 and we are using the observation that a map $1 \to X$ of sets is essentially the same thing as an element of X.

(You may be more familiar with a definition of group in which only the multiplication and perhaps the identity are specified as pieces of *structure*, with the existence of inverses required as a *property*. In that approach, the definition is swiftly followed by a lemma on uniqueness of inverses, guaranteeing that it makes sense to speak of *the* inverse of an element. The two approaches are equivalent, but for many purposes, it is better to frame the definition in the way described in the previous paragraph.)

An algebraic theory consists of two things: first, a collection of operations, each with a specified arity (number of inputs), and second, a collection of equations. For example, the theory of groups has one operation of arity 2, one of arity 1, and one of arity 0. An **algebra** or **model** for an algebraic theory consists of a set X together with a specified map $X^n \to X$ for each operation of arity n, such that the equations hold everywhere. For example, an algebra for the theory of groups is exactly a group.

A more subtle example is the theory of vector spaces over $\mathbb{R}$. This is an algebraic theory with, among other things, an infinite number of operations of arity 1: for each $\lambda \in \mathbb{R}$, we have the operation $\lambda \cdot - : X \to X$ of scalar multiplication by λ (for any vector space X). There is nothing special about the field $\mathbb{R}$ here; the only point is that it was chosen in advance. The theory of vector spaces over $\mathbb{R}$ is different from the theory of vector spaces over $\mathbb{C}$, because they have different operations of arity 1.

In a nutshell, the main property of algebras for an algebraic theory is that the operations are defined everywhere on the set, and the equations hold everywhere too. For example, *every* element of a group has a specified inverse, and *every* element x satisfies the equation $x \cdot x^{-1} = 1$. This is why the theories of groups, rings, and so on, are algebraic theories, but the theory of fields is not.

Example 2.1.5 There are adjunctions

$$
\begin{array}{c}
\mathbf{Top} \\
D \Big\uparrow \dashv \; U \; \dashv \Big\uparrow I \\
\mathbf{Set}
\end{array}
$$

where U sends a space to its set of points, D equips a set with the discrete topology, and I equips a set with the indiscrete topology.

Example 2.1.6 Given sets A and B, we can form their (cartesian) product $A \times B$. We can also form the set B^A of functions from A to B. This is the same as the set $\mathbf{Set}(A, B)$, but we tend to use the notation B^A when we want to emphasize that it is an object of the same category as A and B.

Now fix a set B. Taking the product with B defines a functor

$$
\begin{array}{cccc}
- \times B: & \mathbf{Set} & \to & \mathbf{Set} \\
& A & \mapsto & A \times B.
\end{array}
$$

(Here we are using the blank notation introduced in Example 1.2.12.) There is also a functor

$$
\begin{array}{cccc}
(-)^B: & \mathbf{Set} & \to & \mathbf{Set} \\
& C & \mapsto & C^B.
\end{array}
$$

Moreover, there is a canonical bijection

$$
\mathbf{Set}(A \times B, C) \cong \mathbf{Set}(A, C^B)
$$

for any sets A and C. It is defined by simply changing the punctuation: given a

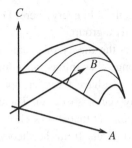

Figure 2.1 In **Set**, a map $A \times B \to C$ can be seen as a way of assigning to each element of A a map $B \to C$.

map $g \colon A \times B \to C$, define $\bar{g} \colon A \to C^B$ by

$$(\bar{g}(a))(b) = g(a, b)$$

($a \in A$, $b \in B$), and in the other direction, given $f \colon A \to C^B$, define $\bar{f} \colon A \times B \to C$ by

$$\bar{f}(a, b) = (f(a))(b)$$

($a \in A$, $b \in B$). Figure 2.1 shows an example with $A = B = C = \mathbb{R}$. By slicing up the surface as shown, a map $\mathbb{R}^2 \to \mathbb{R}$ can be seen as a map from $\mathbb{R}$ to {maps $\mathbb{R} \to \mathbb{R}$}.

Putting all this together, we obtain an adjunction

$$
\begin{array}{c}
\textbf{Set} \\
{\scriptstyle -\times B} \uparrow \dashv \downarrow {\scriptstyle (-)^B} \\
\textbf{Set}
\end{array}
$$

for every set B.

Definition 2.1.7 Let $\mathscr{A}$ be a category. An object $I \in \mathscr{A}$ is **initial** if for every $A \in \mathscr{A}$, there is exactly one map $I \to A$. An object $T \in \mathscr{A}$ is **terminal** if for every $A \in \mathscr{A}$, there is exactly one map $A \to T$.

For example, the empty set is initial in **Set**, the trivial group is initial in **Grp**, and $\mathbb{Z}$ is initial in **Ring** (Example 0.2). The one-element set is terminal in **Set**, the trivial group is terminal (as well as initial) in **Grp**, and the trivial (one-element) ring is terminal in **Ring**. The terminal object of **CAT** is the category **1** containing just one object and one map (necessarily the identity on that object).

A category need not have an initial object, but if it does have one, it is unique up to isomorphism. Indeed, it is unique up to *unique* isomorphism, as follows.

Lemma 2.1.8 *Let I and I' be initial objects of a category. Then there is a unique isomorphism $I \to I'$. In particular, $I \cong I'$.*

Proof Since I is initial, there is a unique map $f \colon I \to I'$. Since I' is initial, there is a unique map $f' \colon I' \to I$. Now $f' \circ f$ and 1_I are both maps $I \to I$, and I is initial, so $f' \circ f = 1_I$. Similarly, $f \circ f' = 1_{I'}$. Hence f is an isomorphism, as required. □

Example 2.1.9 Initial and terminal objects can be described as adjoints. Let $\mathscr{A}$ be a category. There is precisely one functor $\mathscr{A} \to \mathbf{1}$. Also, a functor $\mathbf{1} \to \mathscr{A}$ is essentially just an object of $\mathscr{A}$ (namely, the object to which the unique object of $\mathbf{1}$ is mapped). Viewing functors $\mathbf{1} \to \mathscr{A}$ as objects of $\mathscr{A}$, a left adjoint to $\mathscr{A} \to \mathbf{1}$ is exactly an initial object of $\mathscr{A}$.

Similarly, a right adjoint to the unique functor $\mathscr{A} \to \mathbf{1}$ is exactly a terminal object of $\mathscr{A}$.

Remark 2.1.10 In the language introduced in Remark 1.1.10, the concept of terminal object is dual to the concept of initial object. (More generally, the concepts of left and right adjoint are dual to one another.) Since any two initial objects of a category are uniquely isomorphic, the principle of duality implies that the same is true of terminal objects.

Remark 2.1.11 Adjunctions can be composed. Take adjunctions

$$\mathscr{A} \xrightarrow[\underset{G}{\longleftarrow}]{\overset{F}{\longrightarrow}} {\perp} \ \mathscr{A}' \xrightarrow[\underset{G'}{\longleftarrow}]{\overset{F'}{\longrightarrow}} {\perp} \ \mathscr{A}''$$

where the $\perp$ symbol is a rotated $\dashv$ (thus, $F \dashv G$ and $F' \dashv G'$). Then we obtain an adjunction

$$\mathscr{A} \xrightarrow[\underset{G \circ G'}{\longleftarrow}]{\overset{F' \circ F}{\longrightarrow}} {\perp} \ \mathscr{A}'',$$

since for $A \in \mathscr{A}$ and $A'' \in \mathscr{A}''$,

$$\mathscr{A}''(F'(F(A)), A'') \cong \mathscr{A}'(F(A), G'(A'')) \cong \mathscr{A}(A, G(G'(A'')))$$

naturally in A and A''.

Exercises

2.1.12 Find three examples of adjoint functors not mentioned above. Do the same for initial and terminal objects.

2.1.13 What can be said about adjunctions between discrete categories?

2.1.14 Show that the naturality equations (2.2) and (2.3) can equivalently be replaced by the single equation

$$\left(A' \xrightarrow{p} A \xrightarrow{f} G(B) \xrightarrow{G(q)} G(B')\right) \;=\; \left(F(A') \xrightarrow{F(p)} F(A) \xrightarrow{\bar{f}} B \xrightarrow{q} B'\right)$$

for all p, f and q.

2.1.15 Show that left adjoints preserve initial objects: that is, if $\mathscr{A} \underset{G}{\overset{F}{\rightleftarrows}} \mathscr{B}$ and I is an initial object of $\mathscr{A}$, then $F(I)$ is an initial object of $\mathscr{B}$. Dually, show that right adjoints preserve terminal objects.

(In Section 6.3, we will see this as part of a bigger picture: right adjoints preserve limits and left adjoints preserve colimits.)

2.1.16 Let G be a group.

(a) What interesting functors are there (in either direction) between **Set** and the category $[G, \mathbf{Set}]$ of left G-sets? Which of those functors are adjoint to which?

(b) Similarly, what interesting functors are there between $\mathbf{Vect}_k$ and the category $[G, \mathbf{Vect}_k]$ of k-linear representations of G, and what adjunctions are there between those functors?

2.1.17 Fix a topological space X, and write $\mathscr{O}(X)$ for the poset of open subsets of X, ordered by inclusion. Let

$$\Delta \colon \mathbf{Set} \to [\mathscr{O}(X)^{\mathrm{op}}, \mathbf{Set}]$$

be the functor assigning to a set A the presheaf ΔA with constant value A. Exhibit a chain of adjoint functors

$$\Lambda \dashv \Pi \dashv \Delta \dashv \Gamma \dashv \nabla.$$

2.2 Adjunctions via units and counits

In the previous section, we met the definition of adjunction. In this section and the next, we meet two ways of rephrasing the definition. The one in this section is most useful for theoretical purposes, while the one in the next fits well with many examples.

To start building the theory of adjoint functors, we have to take seriously the naturality requirement (equations (2.2) and (2.3)), which has so far been ignored. Take an adjunction $\mathscr{A} \underset{G}{\overset{F}{\underset{\perp}{\rightleftarrows}}} \mathscr{B}$. Intuitively, naturality says that as A

varies in $\mathscr{A}$ and B varies in $\mathscr{B}$, the isomorphism between $\mathscr{B}(F(A), B)$ and $\mathscr{A}(A, G(B))$ varies in a way that is compatible with all the structure already in place. In other words, it is compatible with composition in the categories $\mathscr{A}$ and $\mathscr{B}$ and the action of the functors F and G.

But what does 'compatible' mean? Suppose, for example, that we have maps

$$F(A) \xrightarrow{g} B \xrightarrow{q} B'$$

in $\mathscr{B}$. There are two things we can do with this data: either compose then take the transpose, which produces a map $\overline{q \circ g} \colon A \to G(B')$, or take the transpose of g then compose it with $G(q)$, which produces a potentially different map $G(q) \circ \bar{g} \colon A \to G(B')$. Compatibility means that they are equal; and that is the first naturality equation (2.2). The second is its dual, and can be explained in a similar way.

For each $A \in \mathscr{A}$, we have a map

$$\left(A \xrightarrow{\eta_A} GF(A) \right) = \left(\overline{F(A) \xrightarrow{1} F(A)} \right).$$

Dually, for each $B \in \mathscr{B}$, we have a map

$$\left(FG(B) \xrightarrow{\varepsilon_B} B \right) = \left(\overline{G(B) \xrightarrow{1} G(B)} \right).$$

(We have begun to omit brackets, writing $GF(A)$ instead of $G(F(A))$, etc.) These define natural transformations

$$\eta \colon 1_{\mathscr{A}} \to G \circ F, \qquad \varepsilon \colon F \circ G \to 1_{\mathscr{B}},$$

called the **unit** and **counit** of the adjunction, respectively.

Example 2.2.1 Take the usual adjunction $\mathbf{Vect}_k \underset{F}{\overset{U}{\underset{\perp}{\rightleftarrows}}} \mathbf{Set}$. Its unit $\eta \colon 1_{\mathbf{Set}} \to U \circ F$ has components

$$\begin{aligned} \eta_S \colon \quad S \quad &\to \quad UF(S) = \{\text{formal } k\text{-linear sums } \textstyle\sum_{s \in S} \lambda_s s\} \\ s \quad &\mapsto \quad s \end{aligned}$$

($S \in \mathbf{Set}$). The component of the counit ε at a vector space V is the linear map

$$\varepsilon_V \colon FU(V) \to V$$

that sends a *formal* linear sum $\sum_{v \in V} \lambda_v v$ to its *actual* value in V.

The vector space $FU(V)$ is enormous. For instance, if $k = \mathbb{R}$ and V is the vector space $\mathbb{R}^2$, then $U(V)$ is the set $\mathbb{R}^2$ and $FU(V)$ is a vector space with one basis element for every element of $\mathbb{R}^2$; thus, it is uncountably infinite-dimensional. Then ε_V is a map from this infinite-dimensional space to the 2-dimensional space V.

Lemma 2.2.2 *Given an adjunction $F \dashv G$ with unit η and counit ε, the triangles*

$$
\begin{array}{ccc}
F \xrightarrow{\ F\eta\ } FGF & \qquad & G \xrightarrow{\ \eta G\ } GFG \\
\ \ \searrow{\scriptstyle 1_F}\ \ \downarrow{\scriptstyle \varepsilon F} & & \ \ \searrow{\scriptstyle 1_G}\ \ \downarrow{\scriptstyle G\varepsilon} \\
F & & G
\end{array}
$$

commute.

Remark 2.2.3 These are called the **triangle identities**. They are commutative diagrams in the functor categories $[\mathscr{A}, \mathscr{B}]$ and $[\mathscr{B}, \mathscr{A}]$, respectively. For an explanation of the notation, see Remarks 1.3.24 (particularly the special cases mentioned on page 37). An equivalent statement is that the triangles

$$
\begin{array}{ccc}
F(A) \xrightarrow{\ F(\eta_A)\ } FGF(A) & \qquad & G(B) \xrightarrow{\ \eta_{G(B)}\ } GFG(B) \\
\ \ \searrow{\scriptstyle 1_{F(A)}}\ \ \downarrow{\scriptstyle \varepsilon_{F(A)}} & & \ \ \searrow{\scriptstyle 1_{G(B)}}\ \ \downarrow{\scriptstyle G(\varepsilon_B)} \\
F(A) & & G(B)
\end{array}
\qquad (2.4)
$$

commute for all $A \in \mathscr{A}$ and $B \in \mathscr{B}$.

Proof of Lemma 2.2.2 We prove that the triangles (2.4) commute. Let $A \in \mathscr{A}$. Since $\overline{1_{GF(A)}} = \varepsilon_{F(A)}$, equation (2.3) gives

$$
\overline{\left(A \xrightarrow{\ \eta_A\ } GF(A) \xrightarrow{\ 1\ } GF(A) \right)} \;=\; \left(F(A) \xrightarrow{\ F(\eta_A)\ } FGF(A) \xrightarrow{\ \varepsilon_{F(A)}\ } F(A) \right).
$$

But the left-hand side is $\overline{\eta_A} = \overline{\overline{1_{F(A)}}} = 1_{F(A)}$, proving the first identity. The second follows by duality.   □

Amazingly, the unit and counit determine the whole adjunction, even though they appear to know only the transposes *of identities*. This is the main content of the following pair of results.

Lemma 2.2.4 *Let $\mathscr{A} \underset{G}{\overset{F}{\rightleftarrows}} \mathscr{B}$ be an adjunction, with unit η and counit ε. Then*

$$
\bar{g} = G(g) \circ \eta_A
$$

for any $g \colon F(A) \to B$, *and*

$$\bar{f} = \varepsilon_B \circ F(f)$$

for any $f \colon A \to G(B)$.

Proof For any map $g \colon F(A) \to B$, we have

$$\overline{\left(F(A) \xrightarrow{g} B\right)} = \overline{\left(F(A) \xrightarrow{1} F(A) \xrightarrow{g} B\right)}$$

$$= \left(A \xrightarrow{\eta_A} GF(A) \xrightarrow{G(g)} G(B)\right)$$

by equation (2.2), giving the first statement. The second follows by duality. $\square$

Theorem 2.2.5 *Take categories and functors* $\mathscr{A} \underset{G}{\overset{F}{\rightleftarrows}} \mathscr{B}$. *There is a one-to-one correspondence between:*

(a) *adjunctions between* F *and* G *(with* F *on the left and* G *on the right);*

(b) *pairs* $\left(1_{\mathscr{A}} \xrightarrow{\eta} GF, FG \xrightarrow{\varepsilon} 1_{\mathscr{B}}\right)$ *of natural transformations satisfying the triangle identities.*

(Recall that by definition, an adjunction between F and G is a choice of isomorphism (2.1) for each A and B, satisfying the naturality equations (2.2) and (2.3).)

Proof We have shown that every adjunction between F and G gives rise to a pair (η, ε) satisfying the triangle identities. We now have to show that this process is bijective. So, take a pair (η, ε) of natural transformations satisfying the triangle identities. We must show that there is a unique adjunction between F and G with unit η and counit ε.

Uniqueness follows from Lemma 2.2.4. For existence, take natural transformations η and ε as in (b). For each A and B, define functions

$$\mathscr{B}(F(A), B) \rightleftarrows \mathscr{A}(A, G(B)), \tag{2.5}$$

both denoted by a bar, as follows. Given $g \in \mathscr{B}(F(A), B)$, put $\bar{g} = G(g) \circ \eta_A \in \mathscr{A}(A, G(B))$. Similarly, in the opposite direction, put $\bar{f} = \varepsilon_B \circ F(f)$.

I claim that for each A and B, the two functions $g \mapsto \bar{g}$ and $f \mapsto \bar{f}$ are mutually inverse. Indeed, given a map $g \colon F(A) \to B$ in $\mathscr{B}$, we have a commutative diagram

$$
\begin{array}{ccc}
F(A) & \xrightarrow{F(\eta_A)} FGF(A) & \xrightarrow{FG(g)} FG(B) \\
& \Big\downarrow{\varepsilon_{F(A)}} & \Big\downarrow{\varepsilon_B} \\
1 \searrow & & \\
F(A) & \xrightarrow[g]{} & B.
\end{array}
$$

The composite map from $F(A)$ to B by one route around the outside of the diagram is

$$\varepsilon_B \circ FG(g) \circ F(\eta_A) = \varepsilon_B \circ F(\bar{g}) = \bar{\bar{g}},$$

and by the other is $g \circ 1 = g$, so $\bar{\bar{g}} = g$. Dually, $\bar{\bar{f}} = f$ for any map $f \colon A \to G(B)$ in $\mathscr{A}$. This proves the claim.

It is straightforward to check the naturality equations (2.2) and (2.3). The functions (2.5) therefore define an adjunction. Finally, its unit and counit are η and ε, since the component of the unit at A is

$$\overline{1_{F(A)}} = G(1_{F(A)}) \circ \eta_A = 1 \circ \eta_A = \eta_A,$$

and dually for the counit. $\square$

Corollary 2.2.6 *Take categories and functors* $\mathscr{A} \underset{G}{\overset{F}{\rightleftarrows}} \mathscr{B}$. *Then* $F \dashv G$ *if and only if there exist natural transformations* $1 \overset{\eta}{\longrightarrow} GF$ *and* $FG \overset{\varepsilon}{\longrightarrow} 1$ *satisfying the triangle identities.* $\square$

Example 2.2.7 An adjunction between ordered sets consists of order-preserving maps $A \underset{g}{\overset{f}{\rightleftarrows}} B$ such that

$$\forall a \in A, \ \forall b \in B, \qquad f(a) \le b \iff a \le g(b). \tag{2.6}$$

This is because both sides of the isomorphism (2.1) in the definition of adjunction are sets with at most one element, so they are isomorphic if and only if they are both empty or both nonempty. The naturality requirements (2.2) and (2.3) hold automatically, since in an ordered set, any two maps with the same domain and codomain are equal.

Recall from Example 1.3.9 that if $C \underset{q}{\overset{p}{\rightrightarrows}} D$ are order-preserving maps of ordered sets then there is at most one natural transformation from p to q, and there is one if and only if $p(c) \le q(c)$ for all $c \in C$. The unit of the adjunction above is the statement that $a \le gf(a)$ for all $a \in A$, and the counit is the statement that $fg(b) \le b$ for all $b \in B$. The triangle identities say nothing, since they assert the equality of two maps in an ordered set with the same domain and codomain.

In the case of ordered sets, Corollary 2.2.6 states that condition (2.6) is equivalent to:

$$\forall a \in A, \ a \le gf(a) \qquad \text{and} \qquad \forall b \in B, \ fg(b) \le b.$$

This equivalence can also be proved directly (Exercise 2.2.10).

For instance, let X be a topological space. Take the set $\mathscr{C}(X)$ of closed subsets of X and the set $\mathscr{P}(X)$ of all subsets of X, both ordered by $\subseteq$. There are order-preserving maps

$$\mathscr{P}(X) \underset{i}{\overset{\text{Cl}}{\rightleftarrows}} \mathscr{C}(X)$$

where i is the inclusion map and Cl is closure. This is an adjunction, with Cl left adjoint to i, as witnessed by the fact that

$$\text{Cl}(A) \subseteq B \iff A \subseteq B$$

for all $A \subseteq X$ and closed $B \subseteq X$. An equivalent statement is that $A \subseteq \text{Cl}(A)$ for all $A \subseteq X$ and $\text{Cl}(B) \subseteq B$ for all closed $B \subseteq X$. Either way, we see that the topological operation of closure arises as an adjoint functor.

Remark 2.2.8   Theorem 2.2.5 states that an adjunction may be regarded as a quadruple $(F, G, \eta, \varepsilon)$ of functors and natural transformations satisfying the triangle identities. An equivalence $(F, G, \eta, \varepsilon)$ of categories (as in Definition 1.3.15) is not necessarily an adjunction. It *is* true that F is left adjoint to G (Exercise 2.3.10), but η and ε are not necessarily the unit and counit (because there is no reason why they should satisfy the triangle identities).

Remark 2.2.9   There is a way of drawing natural transformations that makes the triangle identities intuitively plausible. Suppose, for instance, that we have categories and functors

$$\mathscr{A} \xrightarrow{F_1} \mathscr{C}_1 \xrightarrow{F_2} \mathscr{C}_2 \xrightarrow{F_3} \mathscr{C}_3 \xrightarrow{F_4} \mathscr{B}, \qquad \mathscr{A} \xrightarrow{G_1} \mathscr{D}_1 \xrightarrow{G_2} \mathscr{B}$$

and a natural transformation $\alpha\colon F_4 F_3 F_2 F_1 \to G_2 G_1$. We usually draw α like this:

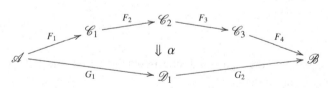

However, we can also draw α as a **string diagram**:

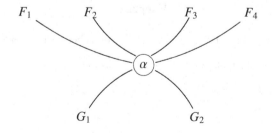

There is nothing special about 4 and 2; we could replace them by any natural numbers m and n. If $m = 0$ then $\mathscr{A} = \mathscr{B}$ and the domain of α is $1_{\mathscr{A}}$ (keeping in mind the last paragraph of Remark 1.1.2(b)). In that case, the disk labelled α has no strings coming into the top. Similarly, if $n = 0$ then there are no strings coming out of the bottom.

Vertical composition of natural transformations corresponds to joining string diagrams together vertically, and horizontal composition corresponds to putting them side by side. The identity on a functor F is drawn as a simple string,

Now let us apply this notation to adjunctions. The unit and counit are drawn as

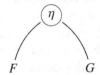 and

The triangle identities now become the topologically plausible equations

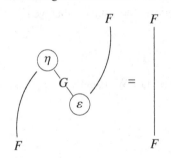 and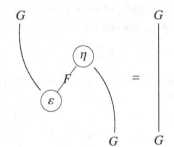

In both equations, the right-hand side is obtained from the left by simply pulling the string straight.

Exercises

2.2.10 Let $A \underset{g}{\overset{f}{\rightleftarrows}} B$ be order-preserving maps between ordered sets. Prove *directly* that the following conditions are equivalent:

(a) for all $a \in A$ and $b \in B$,

$$f(a) \le b \iff a \le g(b);$$

(b) $a \leq g(f(a))$ for all $a \in A$ and $f(g(b)) \leq b$ for all $b \in B$.

(Both conditions state that $f \dashv g$; see Example 2.2.7.)

2.2.11 (a) Let $\mathscr{A} \underset{G}{\overset{F}{\rightleftarrows}} \mathscr{B}$ be an adjunction with unit η and counit ε. Write **Fix**(GF) for the full subcategory of $\mathscr{A}$ whose objects are those $A \in \mathscr{A}$ such that η_A is an isomorphism, and dually **Fix**$(FG) \subseteq \mathscr{B}$. Prove that the adjunction $(F, G, \eta, \varepsilon)$ restricts to an equivalence $(F', G', \eta', \varepsilon')$ between **Fix**(GF) and **Fix**(FG).

(b) Part (a) shows that every adjunction restricts to an equivalence between full subcategories in a canonical way. Take some examples of adjunctions and work out what this equivalence is.

2.2.12 (a) Show that for any adjunction, the right adjoint is full and faithful if and only if the counit is an isomorphism.

(b) An adjunction satisfying the equivalent conditions of part (a) is called a **reflection**. (Compare Example 2.1.3(d).) Of the examples of adjunctions given in this chapter, which are reflections?

2.2.13 (a) Let $f \colon K \to L$ be a map of sets, and denote by $f^* \colon \mathscr{P}(L) \to \mathscr{P}(K)$ the map sending a subset S of L to its inverse image $f^{-1}S \subseteq K$. Then f^* is order-preserving with respect to the inclusion orderings on $\mathscr{P}(K)$ and $\mathscr{P}(L)$, and so can be seen as a functor. Find left and right adjoints to f^*.

(b) Now let X and Y be sets, and write $p \colon X \times Y \to X$ for first projection. Regard a subset S of X as a predicate $S(x)$ in one variable $x \in X$, and similarly a subset R of $X \times Y$ as a predicate $R(x, y)$ in two variables. What, in terms of predicates, are the left and right adjoints to p^*? For each of the adjunctions, interpret the unit and counit as logical implications. (Hint: the left adjoint to p^* is often written as $\exists_Y$, and the right adjoint as $\forall_Y$.)

2.2.14 Given a functor $F \colon \mathscr{A} \to \mathscr{B}$ and a category $\mathscr{S}$, there is a functor $F^* \colon [\mathscr{B}, \mathscr{S}] \to [\mathscr{A}, \mathscr{S}]$ defined on objects $Y \in [\mathscr{B}, \mathscr{S}]$ by $F^*(Y) = Y \circ F$ and on maps α by $F^*(\alpha) = \alpha F$. Show that any adjunction $\mathscr{A} \underset{G}{\overset{F}{\rightleftarrows}} \mathscr{B}$ and category $\mathscr{S}$ give rise to an adjunction

$$[\mathscr{A}, \mathscr{S}] \underset{F^*}{\overset{G^*}{\rightleftarrows}} [\mathscr{B}, \mathscr{S}].$$

(Hint: use Theorem 2.2.5.)

2.3 Adjunctions via initial objects

We now come to the third formulation of adjointness, which is the one you will probably see most often in everyday mathematics.

Consider once more the adjunction

$$\mathbf{Vect}_k$$
$$F \dashv U$$
$$\mathbf{Set}.$$

Let S be a set. The universal property of $F(S)$, the vector space whose basis is S, is most commonly stated like this:

> given a vector space V, any function $f \colon S \to V$ extends uniquely to a linear map $\bar{f} \colon F(S) \to V$.

As remarked in Example 2.1.3(a), forgetful functors are often forgotten: in this statement, '$f \colon S \to V$' should strictly speaking be '$f \colon S \to U(V)$'. Also, the word 'extends' refers implicitly to the embedding

$$\eta_S \colon \quad S \quad \to \quad UF(S)$$
$$\quad s \quad \mapsto \quad s.$$

So in precise language, the statement reads:

> for any $V \in \mathbf{Vect}_k$ and $f \in \mathbf{Set}(S, U(V))$, there is a unique $\bar{f} \in$ $\mathbf{Vect}_k(F(S), V)$ such that the diagram

$$
\begin{array}{ccc}
S & \xrightarrow{\eta_S} & U(F(S)) \\
 & \searrow{\scriptstyle f} & \downarrow{\scriptstyle U(\bar{f})} \\
 & & U(V)
\end{array}
\tag{2.7}
$$

> commutes.

(Compare Example 0.4.) In this section, we show that this statement is equivalent to the statement that F is left adjoint to U with unit η.

To do this, we need a definition.

Definition 2.3.1 Given categories and functors

$$
\begin{array}{c}
\mathscr{B} \\
\downarrow{\scriptstyle Q} \\
\mathscr{A} \xrightarrow{\;P\;} \mathscr{C},
\end{array}
$$

the **comma category** $(P \Rightarrow Q)$ (often written as $(P \downarrow Q)$) is the category defined as follows:

- objects are triples (A, h, B) with $A \in \mathcal{A}$, $B \in \mathcal{B}$, and $h\colon P(A) \to Q(B)$ in $\mathcal{C}$;
- maps $(A, h, B) \to (A', h', B')$ are pairs $(f\colon A \to A', \ g\colon B \to B')$ of maps such that the square

$$
\begin{array}{ccc}
P(A) & \xrightarrow{\ P(f)\ } & P(A') \\
{\scriptstyle h}\downarrow & & \downarrow{\scriptstyle h'} \\
Q(B) & \xrightarrow[\ Q(g)\ ]{} & Q(B')
\end{array}
$$

commutes.

Remark 2.3.2 Given $\mathcal{A}$, $\mathcal{B}$, $\mathcal{C}$, P and Q as above, there are canonical functors and a canonical natural transformation as shown:

$$
\begin{array}{ccc}
(P \Rightarrow Q) & \longrightarrow & \mathcal{B} \\
\downarrow & \nearrow & \downarrow{\scriptstyle Q} \\
\mathcal{A} & \xrightarrow[\ P\ ]{} & \mathcal{C}
\end{array}
$$

In a suitable 2-categorical sense, $(P \Rightarrow Q)$ is universal with this property.

Example 2.3.3 Let $\mathcal{A}$ be a category and $A \in \mathcal{A}$. The **slice category** of $\mathcal{A}$ over A, denoted by $\mathcal{A}/A$, is the category whose objects are maps into A and whose maps are commutative triangles. More precisely, an object is a pair (X, h) with $X \in \mathcal{A}$ and $h\colon X \to A$ in $\mathcal{A}$, and a map $(X, h) \to (X', h')$ in $\mathcal{A}/A$ is a map $f\colon X \to X'$ in $\mathcal{A}$ making the triangle

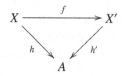

commute.

Slice categories are a special case of comma categories. Recall from Example 2.1.9 that functors $\mathbf{1} \to \mathcal{A}$ are just objects of $\mathcal{A}$. Now, given an object A of $\mathcal{A}$, consider the comma category $(1_{\mathcal{A}} \Rightarrow A)$, as in the diagram

$$
\begin{array}{ccc}
& \mathbf{1} & \\
& \downarrow{\scriptstyle A} & \\
\mathcal{A} & \xrightarrow[\ 1_{\mathcal{A}}\ ]{} & \mathcal{A}.
\end{array}
$$

An object of $(1_{\mathscr{A}} \Rightarrow A)$ is in principle a triple (X, h, B) with $X \in \mathscr{A}$, $B \in \mathbf{1}$, and $h\colon X \to A$ in $\mathscr{A}$; but $\mathbf{1}$ has only one object, so it is essentially just a pair (X, h). Hence the comma category $(1_{\mathscr{A}} \Rightarrow A)$ has the same objects as the slice category $\mathscr{A}/A$. One can check that it has the same maps too, so that $\mathscr{A}/A \cong (1_{\mathscr{A}} \Rightarrow A)$.

Dually (reversing all the arrows), there is a **coslice category** $A/\mathscr{A} \cong (A \Rightarrow 1_{\mathscr{A}})$, whose objects are the maps out of A.

Example 2.3.4   Let $G\colon \mathscr{B} \to \mathscr{A}$ be a functor and let $A \in \mathscr{A}$. We can form the comma category $(A \Rightarrow G)$, as in the diagram

$$
\begin{array}{c}
\mathscr{B} \\
\downarrow{\scriptstyle G} \\
\mathbf{1} \xrightarrow[A]{} \mathscr{A}.
\end{array}
$$

Its objects are pairs $(B \in \mathscr{B}, f\colon A \to G(B))$. A map $(B, f) \to (B', f')$ in $(A \Rightarrow G)$ is a map $q\colon B \to B'$ in $\mathscr{B}$ making the triangle

$$
\begin{array}{ccc}
A & \xrightarrow{\;f\;} & G(B) \\
 & {\scriptstyle f'}\searrow & \downarrow{\scriptstyle G(q)} \\
 & & G(B')
\end{array}
$$

commute.

Notice how this diagram resembles the diagram (2.7) in the vector space example. We will use comma categories $(A \Rightarrow G)$ to capture the kind of universal property discussed there.

Speaking casually, we say that $f\colon A \to G(B)$ is an object of $(A \Rightarrow G)$, when what we should really say is that the pair (B, f) is an object of $(A \Rightarrow G)$. There is potential for confusion here, since there may be different objects B, B' of $\mathscr{B}$ with $G(B) = G(B')$. Nevertheless, we will often use this convention.

We now make the connection between comma categories and adjunctions.

Lemma 2.3.5   *Take an adjunction* $\mathscr{A} \underset{G}{\overset{F}{\underset{\perp}{\rightleftarrows}}} \mathscr{B}$ *and an object* $A \in \mathscr{A}$. *Then the unit map* $\eta_A\colon A \to GF(A)$ *is an initial object of* $(A \Rightarrow G)$.

Proof   Let $(B, f\colon A \to G(B))$ be an object of $(A \Rightarrow G)$. We have to show that there is exactly one map from $(F(A), \eta_A)$ to (B, f).

A map $(F(A), \eta_A) \to (B, f)$ in $(A \Rightarrow G)$ is a map $q\colon F(A) \to B$ in $\mathscr{B}$ such that

$$A \xrightarrow{\eta_A} GF(A)$$
$$\searrow_{f} \quad \downarrow_{G(q)} \quad\quad (2.8)$$
$$G(B)$$

commutes. But $G(q) \circ \eta_A = \bar{q}$ by Lemma 2.2.4, so (2.8) commutes if and only if $f = \bar{q}$, if and only if $q = \bar{f}$. Hence $\bar{f}$ is the unique map $(F(A), \eta_A) \to (B, f)$ in $(A \Rightarrow G)$. □

We now meet our third and final formulation of adjointness.

Theorem 2.3.6 *Take categories and functors $\mathscr{A} \underset{G}{\overset{F}{\rightleftarrows}} \mathscr{B}$. There is a one-to-one correspondence between:*

(a) *adjunctions between F and G (with F on the left and G on the right);*
(b) *natural transformations $\eta: 1_{\mathscr{A}} \to GF$ such that $\eta_A: A \to GF(A)$ is initial in $(A \Rightarrow G)$ for every $A \in \mathscr{A}$.*

Proof We have just shown that every adjunction between F and G gives rise to a natural transformation η with the property stated in (b). To prove the theorem, we have to show that every η with the property in (b) is the unit of exactly one adjunction between F and G.

By Theorem 2.2.5, an adjunction between F and G amounts to a pair (η, ε) of natural transformations satisfying the triangle identities. So it is enough to prove that for every η with the property in (b), there exists a unique natural transformation $\varepsilon: FG \to 1_{\mathscr{B}}$ such that the pair (η, ε) satisfies the triangle identities.

Let $\eta: 1_{\mathscr{A}} \to GF$ be a natural transformation with the property in (b).

Uniqueness Suppose that $\varepsilon, \varepsilon': FG \to 1_{\mathscr{B}}$ are natural transformations such that both (η, ε) and (η, ε') satisfy the triangle identities. One of the triangle identities states that for all $B \in \mathscr{B}$, the triangle

$$G(B) \xrightarrow{\eta_{G(B)}} G(FG(B))$$
$$\searrow_{1} \quad \downarrow_{G(\varepsilon_B)} \quad\quad (2.9)$$
$$G(B)$$

commutes. Thus, ε_B is a map

$$\left(FG(B), \ G(B) \xrightarrow{\eta_{G(B)}} G(FG(B))\right) \quad \longrightarrow \quad \left(B, \ G(B) \xrightarrow{1} G(B)\right)$$

in $(G(B) \Rightarrow G)$. The same is true of ε'_B. But $\eta_{G(B)}$ is initial, so there is only one such map, so $\varepsilon_B = \varepsilon'_B$. This holds for all B, so $\varepsilon = \varepsilon'$.

Existence For $B \in \mathscr{B}$, define $\varepsilon_B \colon FG(B) \to B$ to be the unique map

$$(FG(B), \eta_{G(B)}) \to (B, 1_{G(B)})$$

in $(G(B) \Rightarrow G)$. (So by definition of ε_B, triangle (2.9) commutes.) We show that $(\varepsilon_B)_{B \in \mathscr{B}}$ is a natural transformation $FG \to 1$ such that η and ε satisfy the triangle identities.

To prove naturality, take $B \xrightarrow{q} B'$ in $\mathscr{B}$. We have commutative diagrams

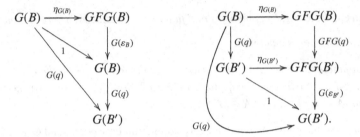

So $q \circ \varepsilon_B$ and $\varepsilon_{B'} \circ FG(q)$ are both maps $\eta_{G(B)} \to G(q)$ in $(G(B) \Rightarrow G)$, and since $\eta_{G(B)}$ is initial, they must be equal. This proves naturality of ε with respect to q. Hence ε is a natural transformation.

We have already observed that one of the triangle identities, equation (2.9), holds. The other states that for $A \in \mathscr{A}$,

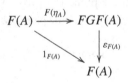

commutes. To prove it, we repeat our previous technique: there are commutative diagrams

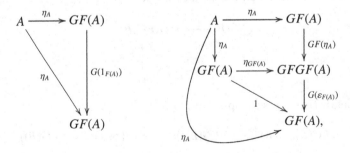

so by initiality of η_A, we have $\varepsilon_{F(A)} \circ F(\eta_A) = 1_{F(A)}$, as required. □

In Section 6.3 we will meet the adjoint functor theorems, which state conditions under which a functor is guaranteed to have a left adjoint. The following corollary is the starting point for their proofs.

Corollary 2.3.7 *Let* $G \colon \mathscr{B} \to \mathscr{A}$ *be a functor. Then* G *has a left adjoint if and only if for each* $A \in \mathscr{A}$, *the category* $(A \Rightarrow G)$ *has an initial object.*

Proof Lemma 2.3.5 proves 'only if'. To prove 'if', let us choose for each $A \in \mathscr{A}$ an initial object of $(A \Rightarrow G)$ and call it $(F(A), \eta_A \colon A \to GF(A))$. (Here $F(A)$ and η_A are just the names we choose to use.) For each map $f \colon A \to A'$ in $\mathscr{A}$, let $F(f) \colon F(A) \to F(A')$ be the unique map such that

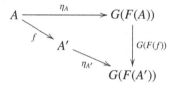

commutes (in other words, the unique map $\eta_A \to \eta_{A'} \circ f$ in $(A \Rightarrow G)$). It is easily checked that F is a functor $\mathscr{A} \to \mathscr{B}$, and the diagram tells us that η is a natural transformation $1 \to GF$. So by Theorem 2.3.6, F is left adjoint to G.□

This corollary justifies the claim made at the beginning of the section: that given functors F and G, to have an adjunction $F \dashv G$ amounts to having maps $\eta_A \colon A \to GF(A)$ with the universal property stated there.

Exercises

2.3.8 What can be said about adjunctions between groups (regarded as one-object categories)?

2.3.9 State the dual of Corollary 2.3.7. How would you prove your dual statement?

2.3.10 Let $(F, G, \eta, \varepsilon)$ be an equivalence of categories, as in Definition 1.3.15. Prove that F is left adjoint to G (heeding the warning in Remark 2.2.8).

2.3.11 Let $\mathscr{A} \underset{F}{\overset{U}{\underset{\top}{\rightleftarrows}}} \mathbf{Set}$ be an adjunction. Suppose that for at least one $A \in \mathscr{A}$, the set $U(A)$ has at least two elements. Prove that for each set S, the unit map $\eta_S \colon S \to UF(S)$ is injective. What does this mean in the case of the usual adjunction between **Grp** and **Set**?

2.3.12 Given sets A and B, a **partial function** from A to B is a pair (S, f) consisting of a subset $S \subseteq A$ and a function $S \to B$. (Think of it as like a function from A to B, but undefined at certain elements of A.) Let **Par** be the category of sets and partial functions.

Show that **Par** is equivalent to **Set**$_*$, the category of sets equipped with a distinguished element and functions preserving distinguished elements. Show also that **Set**$_*$ can be described as a coslice category in a simple way.

3

Interlude on sets

Sets and functions are ubiquitous in mathematics. You might have the impression that they are most strongly connected with the pure end of the subject, but this is an illusion: think of probability density functions in statistics, data sets in experimental science, planetary motion in astronomy, or flow in fluid dynamics.

Category theory is often used to shed light on common constructions and patterns in mathematics. If we hope to do this in an advanced context, we must begin by settling the basic notions of set and function. That is the purpose of the first section of this chapter.

The definition of category mentions a 'collection' of objects and 'collections' of maps. We will see in the second section that some collections are too big to be sets, which leads to a distinction between 'small' and 'large' collections. This distinction will be needed later, most prominently for the adjoint functor theorems (Chapter 6).

The final section takes a historical look at set theory. It also explains why the approach to sets taken in this chapter is more relevant to most of mathematics than the traditional approach is. None of this section is logically necessary for anything that follows, but it may provide useful perspective.

I do not assume that you have encountered axiomatic set theory of any kind. If you have, it is probably best to put it out of your mind while reading this chapter, as the approach to set theory that we take is quite different from the approach that you are most likely to be familiar with. A brief comparison of the traditional and categorical approaches can be found at the very end of the chapter.

3.1 Constructions with sets

We have made no definition of 'set', nor of 'function'. Nevertheless, guided by our intuition, we can list some properties that we expect the world of sets and functions to have. For instance, we can describe some of the sets that we think ought to exist, and some ways of building new sets from old.

Intuitively, a set is a bag of points:

(There may, of course, be infinitely many.) These points, or elements, are not related to one another in any way. They are not in any order, they do not come with any algebraic structure (for instance, there is no specified way of multiplying elements together), and there is no sense of what it means for one point to be close to another. In particular examples, we might have some extra structure in mind; for instance, we often equip the set of real numbers with an order, a field structure and a metric. But to view $\mathbb{R}$ as a mere *set* is to ignore all that structure, to regard it as no more than a bunch of featureless points.

Intuitively, a function $A \to B$ is an assignment of a point in bag B to each point in bag A:

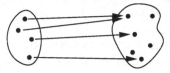

We can do one function after another: given functions

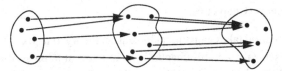

we obtain a composite function

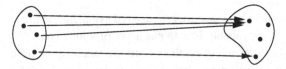

This composition of functions is associative: $h \circ (g \circ f) = (h \circ g) \circ f$. There is also an identity function on every set. Hence:

Sets and functions form a category, denoted by **Set**.

This does not pin things down much: there are many categories, mostly quite unlike the category of sets. So, let us list some of the special features of the category of sets.

The empty set There is a set $\emptyset$ with no elements.

Suppose that someone hands you a pair of sets, A and B, and tells you to specify a function from A to B. Then your task is to specify for each element of A an element of B. The larger A is, the longer the task; the smaller A is, the shorter the task. In particular, if A is empty then the task takes no time at all; we have nothing to do. So there is a function from $\emptyset$ to B specified by doing nothing. On the other hand, there cannot be two different ways to do nothing, so there is only one function from $\emptyset$ to B. Hence:

$\emptyset$ *is an initial object of* **Set**.

In case this argument seems unconvincing, here is an alternative. Suppose that we have a set A with disjoint subsets A_1 and A_2 such that $A_1 \cup A_2 = A$. Then a function from A to B amounts to a function from A_1 to B together with a function from A_2 to B. So if all the sets are finite, we should have the rule

(number of functions from A to B) = (number of functions from A_1 to B)

$$\times \text{(number of functions from } A_2 \text{ to } B).$$

In particular, we could take $A_1 = A$ and $A_2 = \emptyset$. This would force the number of functions from $\emptyset$ to B to be 1. So if we want this rule to hold (and surely we do!), we had better say that there is exactly one function from $\emptyset$ to B.

What about functions *into* $\emptyset$? There is exactly one function $\emptyset \to \emptyset$, namely, the identity. This is a special case of the initiality of $\emptyset$. On the other hand, for a set A that is not empty, there are no functions $A \to \emptyset$, because there is nowhere for elements of A to go.

The one-element set There is a set 1 with exactly one element.

For any set A, there is exactly one function from A to 1, since every element of A must be mapped to the unique element of 1. That is:

1 *is a terminal object of* **Set**.

A function *from* 1 *to* a set B is just a choice of an element of B. In short, the functions $1 \to B$ are the elements of B. Hence:

The concept of element is a special case of the concept of function.

Products Any two sets A and B have a product, $A \times B$. Its elements are the ordered pairs (a, b) with $a \in A$ and $b \in B$. Ordered pairs are familiar from coordinate geometry. All that matters about them is that for $a, a' \in A$ and $b, b' \in B$,

$$(a, b) = (a', b') \iff a = a' \text{ and } b = b'.$$

More generally, take any set I and any family $(A_i)_{i \in I}$ of sets. There is a product set $\prod_{i \in I} A_i$, whose elements are families $(a_i)_{i \in I}$ with $a_i \in A_i$ for each i. Just as for ordered pairs,

$$(a_i)_{i \in I} = (a'_i)_{i \in I} \iff a_i = a'_i \text{ for all } i \in I.$$

Sums Any two sets A and B have a **sum** $A + B$.

Thinking of sets as bags of points, the sum of two sets is obtained by putting all the points into one big bag:

If A and B are finite sets with m and n elements respectively, then $A + B$ always has $m + n$ elements. It makes no difference what the elements of $A + B$ are called; as usual, we only care what $A + B$ is up to isomorphism.

There are inclusion functions

$$A \xrightarrow{\ i\ } A + B \xleftarrow{\ j\ } B$$

such that the union of the images of i and j is all of $A + B$ and the intersection of the images is empty.

Sum is sometimes called **disjoint union** and written as $\amalg$. It is not to be confused with (ordinary) union $\cup$. For a start, we can take the sum of *any* two sets A and B, whereas $A \cup B$ only really makes sense when A and B come as subsets of some larger set. (For to say what $A \cup B$ is, we need to know which elements of A are equal to which elements of B.) And even if A and B do come as subsets of some larger set, $A + B$ and $A \cup B$ can be different. For example, take the subsets $A = \{1, 2, 3\}$ and $B = \{3, 4\}$ of $\mathbb{N}$. Then $A \cup B$ has 4 elements, but $A + B$ has $3 + 2 = 5$ elements.

More generally, any family $(A_i)_{i \in I}$ of sets has a sum $\sum_{i \in I} A_i$. If I is finite and each A_i is finite, say with m_i elements, then $\sum_{i \in I} A_i$ has $\sum_{i \in I} m_i$ elements.

Sets of functions For any two sets A and B, we can form the set A^B of functions from B to A.

This is a special case of the product construction: A^B is the product $\prod_{b \in B} A$ of the constant family $(A)_{b \in B}$. Indeed, an element of $\prod_{b \in B} A$ is a family $(a_b)_{b \in B}$ consisting of one element $a_b \in A$ for each $b \in B$; in other words, it is a function $B \to A$.

Digression on arithmetic We are using notation reminiscent of arithmetic: $A \times B$, $A + B$, and A^B. There is good reason for this: if A is a finite set with m elements and B a finite set with n elements, then $A \times B$ has $m \times n$ elements, $A + B$ has $m + n$ elements, and A^B has m^n elements. Our notation 1 for a one-element set and the alternative notation 0 for the empty set $\emptyset$ also follow this pattern.

All the usual laws of arithmetic have their counterparts for sets:

$$A \times (B + C) \cong (A \times B) + (A \times C),$$
$$A^{B+C} \cong A^B \times A^C,$$
$$(A^B)^C \cong A^{B \times C},$$

and so on, where $\cong$ is isomorphism in the category of sets. (For the last one, see Example 2.1.6.) These isomorphisms hold for all sets, not just finite ones.

The two-element set Let 2 be the set $1 + 1$ (a set with two elements!). For reasons that will soon become clear, I will write the elements of 2 as `true` and `false`.

Let A be a set. Given a subset S of A, we obtain a function $\chi_S : A \to 2$ (the **characteristic function** of $S \subseteq A$), where

$$\chi_S(a) = \begin{cases} \texttt{true} & \text{if } a \in S, \\ \texttt{false} & \text{if } a \notin S \end{cases}$$

$(a \in A)$. Conversely, given a function $f : A \to 2$, we obtain a subset

$$f^{-1}\{\texttt{true}\} = \{a \in A \mid f(a) = \texttt{true}\}$$

of A. These two processes are mutually inverse; that is, χ_S is the unique function $f : A \to 2$ such that $f^{-1}\{\texttt{true}\} = S$. Hence:

Subsets of A correspond one-to-one with functions $A \to 2$.

We already know that the functions from A to 2 form a set, 2^A. When we are thinking of 2^A as the set of all subsets of A, we call it the **power set** of A and write it as $\mathscr{P}(A)$.

Equalizers It would be nice if, given a set A, we could define a subset S of A by specifying a property that the elements of S are to satisfy:

$$S = \{a \in A \mid \text{some property of } a \text{ holds}\}.$$

It is hard to give a general definition of 'property'. There is, however, a special type of property that is easy to handle: equality of two functions. Precisely, given sets and functions $A \overset{f}{\underset{g}{\rightrightarrows}} B$, there is a set

$$\{a \in A \mid f(a) = g(a)\}.$$

This set is called the **equalizer** of f and g, since it is the part of A on which the two functions are equal.

Quotients You are probably familiar with quotient groups and quotient rings (sometimes called factor groups and factor rings) in algebra. Quotients also come up everywhere in topology, such as when we glue together opposite sides of a square to make a cylinder. But the most basic context for quotients is that of sets.

Let A be a set and $\sim$ an equivalence relation on A. There is a set $A/\sim$, the **quotient of A by** $\sim$, whose elements are the equivalence classes. For example, given a group G and a normal subgroup N, define an equivalence relation $\sim$ on G by $g \sim h \iff gh^{-1} \in N$; then $G/\sim = G/N$.

There is also a canonical map

$$p : A \to A/\sim,$$

sending an element of A to its equivalence class. It is surjective, and has the property that $p(a) = p(a') \iff a \sim a'$. In fact, it has a universal property: any function $f : A \to B$ such that

$$\forall a, a' \in A, \qquad a \sim a' \implies f(a) = f(a') \tag{3.1}$$

factorizes uniquely through p, as in the diagram

Thus, for any set B, the functions $A/\sim \to B$ correspond one-to-one with the functions $f : A \to B$ satisfying (3.1). This fact is at the heart of the famous isomorphism theorems of algebra.

We have now listed the properties of sets and functions that will be most important for us. Here are two more.

Natural numbers A function with domain $\mathbb{N}$ is usually called a **sequence**. A crucial property of $\mathbb{N}$ is that sequences can be defined recursively: given a set X, an element $a \in X$, and a function $r: X \to X$, there is a unique sequence $(x_n)_{n=0}^{\infty}$ of elements of X such that

$$x_0 = a, \qquad x_{n+1} = r(x_n) \text{ for all } n \in \mathbb{N}.$$

This property refers to two pieces of structure on $\mathbb{N}$: the element 0, and the function $s: \mathbb{N} \to \mathbb{N}$ defined by $s(n) = n+1$. Reformulated in terms of functions, and writing $x_n = x(n)$, the property is this: for any set X, element $a \in X$, and function $r: X \to X$, there is a unique function $x: \mathbb{N} \to X$ such that $x(0) = 0$ and $x \circ s = r \circ x$. Exercise 3.1.2 asks you to show that this is a *universal* property of $\mathbb{N}$, 0 and s.

Choice Let $f: A \to B$ be a map in a category $\mathscr{A}$. A **section** (or **right inverse**) of f is a map $i: B \to A$ in $\mathscr{A}$ such that $f \circ i = 1_B$.

In the category of sets, any map with a section is certainly surjective. The converse statement is called the **axiom of choice**:

Every surjection has a section.

It is called 'choice' because specifying a section of $f: A \to B$ amounts to choosing, for each $b \in B$, an element of the nonempty set $\{a \in A \mid f(a) = b\}$.

The properties listed above are not theorems, since we do not have rigorous definitions of set and function. What, then, is their status?

Definitions in mathematics usually depend on previous definitions. A vector space is defined as an abelian group with a scalar multiplication. An abelian group is defined as a group with a certain property. A group is defined as a set with certain extra structure. A set is defined as... well, what?

We cannot keep going back indefinitely, otherwise we quite literally would not know what we were talking about. We have to start somewhere. In other words, there have to be some basic concepts not defined in terms of anything else. The concept of set is usually taken to be one of the basic ones, which is why you have probably never read a sentence beginning 'Definition: A **set** is...'. We will treat function as a basic concept, too.

But now there seems to be a problem. If these basic concepts are not defined in terms of anything else, how are we to know what they really are? How are we going to reason in the watertight, logical way upon which mathematics depends? We cannot simply trust our intuitions, since your intuitive idea of set might be slightly different from mine, and if it came to a dispute about how sets behave, we would have no way of deciding who was right.

The problem is solved as follows. Instead of *defining* a set to be a such-and-such and a function to be a such-and-such else, we list some *properties* that we assume sets and functions to have. In other words, we never attempt to say what sets and functions *are*; we just say what you can *do* with them.

In his excellent book *Mathematics: A Very Short Introduction*, Timothy Gowers (2002) considers the question: 'What is the black king in chess?' He swiftly points out that this question is rather peculiar. It is not important that the black king *is* a small piece of wood, painted a certain colour and carved into a certain shape. We could equally well use a scrap of paper with 'BK' written on it. What matters is what the black king *does*: it can move in certain ways but not others, according to the rules of chess.

Similarly, we might not be able to say directly what a set or function 'is', but we agree that they are to satisfy all the properties on the list. So the list of properties acts as an agreement on how to use the words 'set' and 'function', just as the rules of chess act as an agreement on how to use the chess pieces.

What we are doing is often referred to as *foundations*. In this metaphor, the foundation consists of the basic concepts (set and function), which are not built on anything else, but are assumed to satisfy a stated list of properties. On top of the foundations are built some basic definitions and theorems. On top of those are built further definitions and theorems, and so on, towering upwards.

The properties above are stated informally, but they can be formalized using some categorical language. (See Lawvere and Rosebrugh (2003) or Leinster (2014).) In the formal version, we begin by saying that sets and functions form a category, **Set**. We then list some properties of this category. For example, the category is required to have an initial and a terminal object, and the properties described informally under the headings 'Products' and 'Equalizers' are made formal by the statement that **Set** 'has limits' (a phrase defined in Chapter 5).

While we were making the list, we were guided by our intuition about sets. But once it is made, our intuition plays no further official role: any disputes about the nature of sets are settled by consulting the list of properties.

(A subtlety arises. Whatever list of properties one writes down, there might be some questions that cannot be settled. In other words, there might be multiple inequivalent categories satisfying all the properties listed. This gets us into the realm of advanced logic: Gödel incompleteness, the continuum hypothesis, and so on, all beyond the scope of this book.)

Now let us look again at the section on the empty set. You might have felt that I was on shaky ground when trying to convince you that ∅ is initial. But the point is that I do not need to convince you that this is a *true statement*; I only need to convince you that it is a *convenient assumption*. Compare the rule for numbers that $x^0 = 1$. One can reasonably argue that 0 copies of x multiplied together ought to be 1, but really the best justification for this rule is convenience: it makes other rules such as $x^{m+n} = x^m \cdot x^n$ true without exception. Indeed, it does not even make sense to ask whether it is 'true' that $\emptyset$ is initial until we have written down our assumptions about how sets and functions behave. For until then, what could 'true' mean? There is no physical world of sets against which to test such statements.

We can make whatever assumptions about sets we like, but some lead to more interesting mathematics than others. If, for instance, you want to assume that there are *no* functions from $\emptyset$ to any other set, you can, but the tower of mathematics built on that foundation will look different from what you are used to, and probably not in a good way. For example, the 'number of functions' rule (page 67) will fail, and there will be further unpleasant surprises higher up the tower.

Exercises

3.1.1 The **diagonal functor** $\Delta\colon \mathbf{Set} \to \mathbf{Set} \times \mathbf{Set}$ is defined by $\Delta(A) = (A, A)$ for all sets A. Exhibit left and right adjoints to Δ.

3.1.2 In the paragraph headed 'Natural numbers', it was observed that the set $\mathbb{N}$, together with the element 0 and the function $s\colon \mathbb{N} \to \mathbb{N}$, has a certain property. This property can be understood as stating that the triple $(\mathbb{N}, 0, s)$ is the initial object of a certain category $\mathscr{C}$. Find $\mathscr{C}$.

3.2 Small and large categories

We have now made some assumptions about the nature of sets. One consequence of those assumptions is that in many of the categories we have met, the collection of all objects is too large to form a set. In fact, even the collection of *isomorphism classes* of objects is often too large to form a set. In this section, I will explain what these statements mean, and prove them.

This section is not of central importance. As this book proceeds, I will say as little as possible about the distinction between sets and collections too large to be sets. Nevertheless, the distinction begins to matter in parts of category theory lying just within the scope of this book (the adjoint functor theorems), as well as beyond.

Given sets A and B, write $|A| \leq |B|$ (or $|B| \geq |A|$) if there exists an injection $A \to B$. We give no meaning to the expression '$|A|$' or '$|B|$' in isolation. (It would perhaps be more logical to write $A \leq B$ rather than $|A| \leq |B|$, but the notation is well-established.) In the case of finite sets, it just means that the number of elements of A is less than or equal to the number of elements of B.

Since identity maps are injective, $|A| \leq |A|$ for all sets A, and since the composite of two injections is an injection,

$$|A| \leq |B| \leq |C| \implies |A| \leq |C|.$$

Also, if $A \cong B$ then $|A| \leq |B| \leq |A|$. Less obvious is the converse:

Theorem 3.2.1 (Cantor–Bernstein)   *Let A and B be sets. If $|A| \leq |B| \leq |A|$ then $A \cong B$.*

Proof   Exercise 3.2.12. $\square$

These observations tell us that $\leq$ is a preorder (Example 1.1.8(e)) on the collection of all sets. It is not a genuine order, since $|A| \leq |B| \leq |A|$ only implies that $A \cong B$, not $A = B$. We write $|A| = |B|$, and say that A and B **have the same cardinality**, if $A \cong B$, or equivalently if $|A| \leq |B| \leq |A|$.

As long as we do not confuse equality with isomorphism, the sign $\leq$ behaves as we might imagine. For example, write $|A| < |B|$ if $|A| \leq |B|$ and $|A| \neq |B|$. Then

$$|A| \leq |B| < |C| \implies |A| < |C| \tag{3.2}$$

for sets A, B and C. Indeed, we have already established that $|A| \leq |C|$, and the strict inequality follows from Theorem 3.2.1.

Here is another fundamental result of set theory.

Theorem 3.2.2 (Cantor)   *Let A be a set. Then $|A| < |\mathscr{P}(A)|$.*

Recall that $\mathscr{P}(A)$ is the power set of A. The lemma is easy for finite sets, since if A has n elements then $\mathscr{P}(A)$ has 2^n elements, and $n < 2^n$.

Proof   Exercise 3.2.13. $\square$

Corollary 3.2.3   *For every set A, there is a set B such that $|A| < |B|$.* $\square$

In other words, there is no biggest set.

We now justify the claim made at the beginning of this section: that for many familiar categories, the collection of isomorphism classes of objects is too large to form a set. We begin by doing this for the category **Set** itself.

As a clue to why the collection of isomorphism classes of sets might be too large to form a set, consider the following statement: the collection of isomorphism classes of *finite* sets is too large to form a *finite* set. This is because there is one isomorphism class of finite sets for each natural number, but there are infinitely many natural numbers.

Proposition 3.2.4 *Let I be a set, and let $(A_i)_{i \in I}$ be a family of sets. Then there exists a set not isomorphic to any of the sets A_i.*

Proof Put

$$
A = \mathscr{P}\left(\sum_{i \in I} A_i\right),
$$

the power set of the sum of the sets A_i. For each $j \in I$, we have the inclusion function $A_j \to \sum_{i \in I} A_i$, so by Theorem 3.2.2,

$$
|A_j| \le \left|\sum_{i \in I} A_i\right| < |A|.
$$

Hence $|A_j| < |A|$ by (3.2), and in particular, $A_j \not\cong A$. □

We use the word **class** informally to mean any collection of mathematical objects. All sets are classes, but some classes (such as the class of all sets) are too big to be sets. A class will be called **small** if it is a set, and **large** otherwise. For example, Proposition 3.2.4 states that the class of isomorphism classes of sets is large. The crucial point is:

Any individual *set is small, but the* class *of sets is large.*

This is even true if we pretend that isomorphic sets are equal.

Although the 'definition' of class is not precise, it will do for our purposes. We make a naive distinction between small and large collections, and implicitly use some intuitively plausible principles (for example, that any subcollection of a small collection is small).

A category $\mathscr{A}$ is **small** if the class or collection of all maps in $\mathscr{A}$ is small, and **large** otherwise. If $\mathscr{A}$ is small then the class of objects of $\mathscr{A}$ is small too, since objects correspond one-to-one with identity maps.

A category $\mathscr{A}$ is **locally small** if for each $A, B \in \mathscr{A}$, the class $\mathscr{A}(A, B)$ is small. (So, small implies locally small.) Many authors take local smallness to be part of the definition of category. The class $\mathscr{A}(A, B)$ is often called the **hom-set** from A to B, although strictly speaking, we should only call it this when $\mathscr{A}$ is locally small.

Example 3.2.5 **Set** is locally small, because for any two sets A and B, the functions from A to B form a set. This was one of the properties of sets stated in Section 3.1.

Example 3.2.6 $\mathbf{Vect}_k$, **Grp**, **Ab**, **Ring** and **Top** are all locally small. For example, given rings A and B, a homomorphism from A to B is a function from A to B with certain properties, and the collection of all functions from A to B is small, so the collection of homomorphisms from A to B is certainly small.

A category is small if and only if it is locally small and its class of objects is small. Again, it may help to consider a similar fact about finiteness: a category $\mathscr{A}$ is finite (that is, the class of all maps in $\mathscr{A}$ is finite) if and only if it is locally finite (that is, each class $\mathscr{A}(A, B)$ is finite) and its class of objects is finite.

Example 3.2.7   Consider the category $\mathscr{B}$ defined in the last paragraph of Example 1.3.20. Its objects correspond to the natural numbers, which form a set, so the class of objects of $\mathscr{B}$ is small. Each hom-set $\mathscr{B}(\mathbf{m}, \mathbf{n})$ is a set (indeed, a finite set), so $\mathscr{B}$ is locally small. Hence $\mathscr{B}$ is small.

A category is **essentially small** if it is equivalent to some small category. For example, the category of finite sets is essentially small, since by Example 1.3.20, it is equivalent to the small category $\mathscr{B}$ just mentioned.

If two categories $\mathscr{A}$ and $\mathscr{B}$ are equivalent, the class of isomorphism classes of objects of $\mathscr{A}$ is in bijection with that of $\mathscr{B}$. In a small category, the class of objects is small, so the class of isomorphism classes of objects is certainly small. Hence in an essentially small category, the class of isomorphism classes of objects is small. From this we deduce:

Proposition 3.2.8   **Set** *is not essentially small.*

Proof   Proposition 3.2.4 states that the class of isomorphism classes of sets is large. The result follows.   □

By adapting this argument, we can show that many of our standard examples of categories are not essentially small. The strategy is to prove that there are at least as many objects of our category as there are sets.

Example 3.2.9   For any field k, the category $\mathbf{Vect}_k$ of vector spaces over k is not essentially small. As in the proof of Proposition 3.2.8, it is enough to prove that the class of isomorphism classes of vector spaces is large. In other words, it is enough to prove that for any set I and family $(V_i)_{i \in I}$ of vector spaces, there exists a vector space not isomorphic to any of the spaces V_i.

To show this, write $\mathbf{Vect}_k \underset{F}{\overset{U}{\underset{\longleftarrow}{\overset{\top}{\longrightarrow}}}} \mathbf{Set}$ for the free and forgetful functors. As in the proof of Proposition 3.2.4, the set

$$S = \mathscr{P}\left(\sum_{i \in I} U(V_i) \right)$$

has the property that $|U(V_i)| < |S|$ for all $i \in I$. The free vector space $F(S)$ on S contains a copy of S as a basis, so $|S| \le |UF(S)|$. Hence $|U(V_i)| < |UF(S)|$ for all i, and so $F(S) \not\cong V_i$ for all i, as required.

Similarly, none of the categories **Grp**, **Ab**, **Ring** and **Top** is essentially small (Exercise 3.2.14).

Recall that the category of *all* categories and functors is written as **CAT**.

Definition 3.2.10 We denote by **Cat** the category of small categories and functors between them.

Example 3.2.11 Monoids are by definition *sets* equipped with certain structure, so the one-object categories that they correspond to are small. Let $\mathscr{M}$ be the full subcategory of **Cat** consisting of the one-object categories. Then there is an equivalence of categories **Mon** $\simeq \mathscr{M}$. This is proved by the argument in Example 1.3.21, noting that because each object of $\mathscr{M}$ is a *small* one-object category, the collection of maps from the single object to itself really is a set.

Exercises

3.2.12 (a) Let A be a set. Let $\theta\colon \mathscr{P}(A) \to \mathscr{P}(A)$ be a map that is order-preserving with respect to inclusion. A **fixed point** of θ is an element $S \in \mathscr{P}(A)$ such that $\theta(S) = S$. By considering

$$S = \bigcup_{R \in \mathscr{P}(A)\,:\, \theta(R) \supseteq R} R,$$

prove that θ has at least one fixed point.

(b) Take sets and functions $A \underset{g}{\overset{f}{\rightleftarrows}} B$. Using (a), show that there is some subset S of A such that $g(B \setminus fS) = A \setminus S$.

(c) Deduce the Cantor–Bernstein theorem (Theorem 3.2.1).

3.2.13 (a) Let A be a set and $f\colon A \to \mathscr{P}(A)$ a function. By considering

$$\{a \in A \mid a \notin f(a)\},$$

prove that f is not surjective.

(b) Deduce Cantor's theorem (Theorem 3.2.2): $|A| < |\mathscr{P}(A)|$ for all sets A.

3.2.14 (a) Let $\mathscr{A}$ be a category. Suppose there exists a functor $U\colon \mathscr{A} \to \textbf{Set}$ such that U has a left adjoint and for at least one $A \in \mathscr{A}$, the set $U(A)$ has at least two elements. Prove that for any set I and any family $(A_i)_{i \in I}$ of objects of $\mathscr{A}$, there is some object of $\mathscr{A}$ not isomorphic to A_i for any $i \in I$. (Hint: use Exercise 2.3.11.)

(b) Let $\mathscr{A}$ be a category satisfying the assumption of (a). Prove that $\mathscr{A}$ is not essentially small.

(c) Deduce that none of the categories **Set**, **Vect**$_k$, **Grp**, **Ab**, **Ring**, and **Top** is essentially small.

3.2.15 Which of the following categories are small? Which are locally small?

(a) **Mon**, the category of monoids;

(b) $\mathbb{Z}$, the group of integers, viewed as a one-object category;

(c) $\mathbb{Z}$, the ordered set of integers;

(d) **Cat**, the category of small categories;

(e) the multiplicative monoid of cardinals.

3.2.16 Let $O\colon$ **Cat** $\to$ **Set** be the functor sending a small category to its set of objects. Exhibit a chain of adjoints $C \dashv D \dashv O \dashv I$.

3.3 Historical remarks

The set theory that we began to develop in Section 3.1 is rather different from what many mathematicians think of as set theory. Here I will explain what the socially dominant version of set theory is, why, despite its dominance, it is the object of widespread suspicion, and why the kind of set theory outlined here is a more accurate reflection of how mathematicians use sets in practice.

Cantor's set theory The creation of set theory is generally credited to the German mathematician Georg Cantor, in the late nineteenth century. Previously, sets had seldom been regarded as entities worthy of study in their own right; but Cantor, originally motivated by a problem in Fourier analysis, developed an extensive theory. Among many other things, he showed that there are different sizes of infinity, proving, for instance, that there is no bijection between $\mathbb{N}$ and $\mathbb{R}$.

Cantor's theory met all the resistance that typically greets a really new idea. His work was criticized as nonsensical, as meaningless, as far too abstract; then later, as all very well but of no use to the mainstream of mathematics. Kronecker, an important mathematician of the day, called him a charlatan and a corrupter of youth. But nowadays, the basics of Cantor's work are on nearly every undergraduate mathematics syllabus.

Times change. In the modern style of mathematics, almost every definition, when unravelled sufficiently, depends on the notion of set. But pre-Cantor, this was not so. It is interesting to try to understand the outlook of mathematicians of the time, who had successfully developed sophisticated subjects such as complex analysis and Galois theory without depending on this notion that we now regard as fundamental.

Before continuing with the history, we need to discuss another fundamental concept.

Types Suppose someone asks you 'is $\sqrt{2} = \pi$?' Your answer is, of course, 'no'. Now suppose someone asks you 'is $\sqrt{2} = \log$?' You might frown and wonder if you had heard right, and perhaps your answer would again be 'no'; but it would be a different kind of 'no'. After all, $\sqrt{2}$ is a number, whereas log is a function, so it is inconceivable that they could be equal. A better answer would be 'your question makes no sense'.

This illustrates the idea of **types**. The square root of 2 is a real number, $\mathbb{Q}$ is a field, S_3 is a group, log is a function from $(0, \infty)$ to $\mathbb{R}$, and $\frac{d}{dx}$ is an operation that takes as input one function from $\mathbb{R}$ to $\mathbb{R}$ and produces as output another such function. One says that the type of $\sqrt{2}$ is 'real number', the type of $\mathbb{Q}$ is 'field', and so on. We all have an inbuilt sense of type, and it would not usually occur to us to ask whether two things of different type were equal.

You may have met this idea before if you have programmed computers. Many programming languages require you to declare the type of a variable before you first use it. For example, you might declare that x is to be a variable of type 'real number', n a variable of type 'integer', M a variable of type '3×3 matrix of lists of binary digits', and so on.

The distinction between different types of object has always been instinctively understood. At the beginning of the twentieth century, however, events took a strange turn.

Membership-based set theory Those who came after Cantor sought to compile a definitive list of assumptions to be made about sets: an *axiomatization* of set theory. The list they arrived at, in the early years of the twentieth century, is known as ZFC (Zermelo–Fraenkel with Choice). It soon became the standard, and it is the only kind of axiomatic set theory that most present-day mathematicians know.

The axiomatization of Zermelo et al. was in some ways similar to the one that we were working towards in the first section of this chapter. But there is at least one crucial difference: whereas we took sets and *functions* as our basic concepts, they took sets and *membership*.

At first sight, this difference might seem mild. But when the membership-based approach is used as a foundation on which to build the rest of mathematics, several bizarre features become apparent:

- In the Zermelo approach, *everything* is a set. For instance, a function is defined as a set with certain properties. Many other things that you would not think of as being sets are, nevertheless, treated as sets: the number $\sqrt{2}$ is a set, the function log is a set, the operator $\frac{d}{dx}$ is a set, and so on.

  You might wonder how this is possible. Perhaps it is useful to compare data storage in a computer, where files of all different types (text, sound, images, and so on) are ultimately encoded as sequences of 0s and 1s. To give an example, in the membership-based set theory presented in most books, the number 4 is encoded as the set

  $$\{\emptyset, \{\emptyset\}, \{\emptyset, \{\emptyset\}\}, \{\emptyset, \{\emptyset\}, \{\emptyset, \{\emptyset\}\}\}\}.$$

- The virtue of this approach is its simplicity: *everything* is a set! But the price to be paid is very high: we lose the fundamental notion of type, precisely because everything is regarded as being of type 'set'.

- In the Zermelo approach, the elements of sets are always sets too. This is in conflict with ordinary mathematics. For instance, in ordinary mathematics, $\mathbb{R}$ is certainly a set, but real numbers themselves are not regarded as sets. (After all, what is an element of π?)

- In this approach, membership is a global relation, meaning that for *any* two sets A and B, it makes sense to ask whether $A \in B$. Since this approach views everything as a set, it makes sense to ask such apparently nonsensical questions as 'is $\mathbb{Q} \in \sqrt{2}$?'

  Further still, the axioms of ZFC imply that we can form the intersection $A \cap B$ of *any* sets A and B. (Its elements are those sets C for which $C \in A$ and $C \in B$.) This makes possible further nonsensical questions such as 'does the cyclic group of order 10 have nonempty intersection with $\mathbb{Z}$?'

  The answers to these nonsensical questions depend on the fine detail of how mathematical objects (numbers, functions, groups, etc.) are encoded as sets. Even devotees of the membership-based approach agree that this encoding is a matter of convention, just like a word processor's encoding of a document as a string of 0s and 1s. So the answers to these questions are meaningless.

Set theory today It should now be apparent why many modern-day mathematicians are suspicious of set theory. However often they are told that it is 'the foundation of mathematics', they feel that much of it is irrelevant to their concerns.

To some extent, this is justified. But it is also a symptom of the historical dominance of membership-based set theory: most mathematicians do not realize that there is any other kind. This is a shame. Taking sets and functions (rather than sets and membership) as the basic concepts leads to a theory containing all of the meaningful results of Cantor and others, but with none of the aspects that seem so remote from the rest of mathematics. In particular, the function-based approach respects the fundamental notion of type.

The function-based approach is, of course, categorical, and its advantages are related to more general points about how mathematics looks through categorical eyes. Objects are understood through their place in the ambient category. We get inside an object by probing it with maps to or from other objects. For example, an element of a set A is a map $1 \to A$, and a subset of A is a map $A \to 2$. Probing of this kind is the main theme of the next chapter.

Footnote for those familiar with ZFC People brought up on traditional axiomatic set theory often have the following concern when they come across categorical set theory for the first time. The objects and maps of a category form a collection of some kind, perhaps a set, so the notion of category appears to depend on some prior set-like notion. How, then, can sets be axiomatized categorically? Is that not circular?

It is not, because sets can be axiomatized categorically without mentioning categories once. To see how, let us first recall the shape of the ZFC axiomatization of sets. Informally, it looks like this:

- there are some things called sets;
- there is a binary relation on sets, called membership ($\in$);
- some axioms hold.

A categorical axiomatization of sets looks, informally, like this:

- there are some things called sets;
- for each set A and set B, there are some things called functions from A to B;
- to each function f from A to B and function g from B to C, there is assigned a function $g \circ f$ from A to C;
- some axioms hold.

Making precise such phrases as 'some things' requires delicacy, as will be familiar to anyone who has done a logic course. But the difficulties are no worse for categorical axiomatizations of sets than for membership-based axiomatizations such as ZFC.

One popular choice of categorical axioms for set theory can be summarized informally as follows.

1. Composition of functions is associative and has identities.
2. There is a terminal set.
3. There is a set with no elements.
4. A function is determined by its effect on elements.
5. Given sets A and B, one can form their product $A \times B$.
6. Given sets A and B, one can form the set of functions from A to B.
7. Given $f \colon A \to B$ and $b \in B$, one can form the inverse image $f^{-1}\{b\}$.
8. The subsets of a set A correspond to the functions from A to $\{0, 1\}$.
9. The natural numbers form a set.
10. Every surjection has a section.

This informal summary uses terms such as 'element' and 'inverse image', which can be defined in terms of the basic concepts of set, function and composition. For instance, an element of a set A is defined as a map from the terminal set to A.

It is certainly *convenient* to express these axioms in terms of categories. For example, the first axiom says that sets and functions form a category, and all ten together can be expressed in categorical jargon as 'sets and functions form a well-pointed topos with natural numbers object and choice'. But in order to state the axioms, it is not *necessary* to appeal to any general notion of category. They can be expressed directly in terms of sets and functions. For details, see Lawvere and Rosebrugh (2003) or Leinster (2014).

Exercise

3.3.1 Choose a mathematician at random. Ask them whether they can accurately state any axiomatization of sets (without looking it up). If not, ask them what operating principles they actually use when handling sets in their day-to-day work.

4

Representatables

A category is a world of objects, all looking at one another. Each sees the world from a different viewpoint.

Consider, for instance, the category of topological spaces, and let us ask how it looks when viewed from the one-point space 1. A map from 1 to a space X is essentially the same thing as a point of X, so we might say that 1 'sees points'. Similarly, a map from $\mathbb{R}$ to a space X could reasonably be called a curve in X, and in this sense, $\mathbb{R}$ sees curves.

Now consider the category of groups. A map from the infinite cyclic group $\mathbb{Z}$ to a group G amounts to an element of G. (For given $g \in G$, there is a unique homomorphism $\phi \colon \mathbb{Z} \to G$ such that $\phi(1) = g$.) So, $\mathbb{Z}$ sees elements. Similarly, if p is a prime number then the cyclic group $\mathbb{Z}/p\mathbb{Z}$ sees elements of order 1 or p.

Any ring homomorphism between fields is injective, so in the category of fields, a map $K \to L$ is a way of realizing L as an extension of K. Hence each field K sees the extensions of itself. If K and L are fields of different characteristic then there are no homomorphisms between K and L, so the category of fields is the union of disjoint subcategories $\mathbf{Field}_0$, $\mathbf{Field}_2$, $\mathbf{Field}_3$, $\mathbf{Field}_5$, ... consisting of the fields of characteristics $0, 2, 3, 5, \ldots$. Each field is blind to the fields of different characteristic.

In the ordered set $(\mathbb{R}, \leq)$, the object 0 sees whether a number is nonnegative. In other words, if x is nonnegative then there is one map $0 \to x$, and if not, there are none.

We can also ask the dual question: fixing an object of a category, what are the maps *into* it? Let S be the two-element set, for instance. For an arbitrary set X, the maps from X to S correspond to the subsets of X (as we saw in Section 3.1). Now give S the topology in which one of the singleton subsets is open but the other is not. For any topological space X, the continuous maps from X into S correspond to the *open* subsets of X.

This chapter explores the theme of how each object sees and is seen by the category in which it lives. We are naturally led to the notion of representable functor, which (after adjunctions) provides our second approach to the idea of universal property.

4.1 Definitions and examples

Fix an object A of a category $\mathscr{A}$. We will consider the totality of maps out of A. To each $B \in \mathscr{A}$, there is assigned the set (or class) $\mathscr{A}(A, B)$ of maps from A to B. The content of the following definition is that this assignation is functorial in B: any map $B \to B'$ induces a function $\mathscr{A}(A, B) \to \mathscr{A}(A, B')$.

Definition 4.1.1 Let $\mathscr{A}$ be a locally small category and $A \in \mathscr{A}$. We define a functor

$$H^A = \mathscr{A}(A, -) \colon \mathscr{A} \to \mathbf{Set}$$

as follows:

- for objects $B \in \mathscr{A}$, put $H^A(B) = \mathscr{A}(A, B)$;
- for maps $B \xrightarrow{g} B'$ in $\mathscr{A}$, define

$$H^A(g) = \mathscr{A}(A, g) \colon \mathscr{A}(A, B) \to \mathscr{A}(A, B')$$

by

$$p \mapsto g \circ p$$

for all $p \colon A \to B$.

Remarks 4.1.2 (a) Recall that 'locally small' means that each class $\mathscr{A}(A, B)$ is in fact a set. This hypothesis is clearly necessary in order for the definition to make sense.

(b) Sometimes $H^A(g)$ is written as $g \circ -$ or g_*. All three forms, as well as $\mathscr{A}(A, g)$, are in use.

Definition 4.1.3 Let $\mathscr{A}$ be a locally small category. A functor $X \colon \mathscr{A} \to \mathbf{Set}$ is **representable** if $X \cong H^A$ for some $A \in \mathscr{A}$. A **representation** of X is a choice of an object $A \in \mathscr{A}$ and an isomorphism between H^A and X.

Representable functors are sometimes just called 'representables'. Only set-valued functors (that is, functors with codomain **Set**) can be representable.

Example 4.1.4 Consider H^1: **Set** $\to$ **Set**, where 1 is the one-element set. Since a map from 1 to a set B amounts to an element of B, we have

$$H^1(B) \cong B$$

for each $B \in$ **Set**. It is easily verified that this isomorphism is natural in B, so H^1 is isomorphic to the identity functor $1_{\textbf{Set}}$. Hence $1_{\textbf{Set}}$ is representable.

Example 4.1.5 All of the 'seeing' functors in the introduction to this chapter are representable. The forgetful functor **Top** $\to$ **Set** is isomorphic to $H^1 =$ **Top**$(1, -)$, and the forgetful functor **Grp** $\to$ **Set** is isomorphic to **Grp**$(\mathbb{Z}, -)$. For each prime p, there is a functor U_p: **Grp** $\to$ **Set** defined on objects by

$$U_p(G) = \{\text{elements of } G \text{ of order 1 or } p\},$$

and as claimed above, $U_p \cong$ **Grp**$(\mathbb{Z}/p\mathbb{Z}, -)$ (Exercise 4.1.28). Hence U_p is representable.

Example 4.1.6 There is a functor ob : **Cat** $\to$ **Set** sending a small category to its set of objects. (The category **Cat** was introduced in Definition 3.2.10.) It is representable. Indeed, consider the terminal category **1** (with one object and only the identity map). A functor from **1** to a category $\mathscr{B}$ simply picks out an object of $\mathscr{B}$. Thus,

$$H^1(\mathscr{B}) \cong \text{ob } \mathscr{B}.$$

Again, it is easily verified that this isomorphism is natural in $\mathscr{B}$; hence ob $\cong$ **Cat**$(1, -)$. It can be shown similarly that the functor **Cat** $\to$ **Set** sending a small category to its set of maps is representable (Exercise 4.1.31).

Example 4.1.7 Let M be a monoid, regarded as a one-object category. Recall from Example 1.2.8 that a set-valued functor on M is just an M-set. Since the category M has only one object, there is only one representable functor on it (up to isomorphism). As an M-set, the unique representable is the so-called **left regular representation** of M, that is, the underlying set of M acted on by multiplication on the left.

Example 4.1.8 Let **Toph**$_*$ be the category whose objects are topological spaces equipped with a basepoint and whose arrows are homotopy classes of basepoint-preserving continuous maps. Let $S^1 \in$ **Toph**$_*$ be the circle. Then for any object $X \in$ **Toph**$_*$, the maps $S^1 \to X$ in **Toph**$_*$ are the elements of the fundamental group $\pi_1(X)$. Formally, this says that the composite functor

$$\textbf{Toph}_* \xrightarrow{\pi_1} \textbf{Grp} \xrightarrow{U} \textbf{Set}$$

is isomorphic to **Toph**$_*(S^1, -)$. In particular, it is representable.

Example 4.1.9   Fix a field k and vector spaces U and V over k. There is a functor

$$\mathbf{Bilin}(U, V; -)\colon \mathbf{Vect}_k \to \mathbf{Set}$$

whose value $\mathbf{Bilin}(U, V; W)$ at $W \in \mathbf{Vect}_k$ is the set of bilinear maps $U \times V \to W$. It can be shown that this functor is representable; in other words, there is a space T with the property that

$$\mathbf{Bilin}(U, V; W) \cong \mathbf{Vect}_k(T, W)$$

naturally in W. This T is the tensor product $U \otimes V$, which we met just after the proof of Lemma 0.7.

Adjunctions give rise to representable functors in the following way.

Lemma 4.1.10   *Let* $\mathscr{A} \underset{G}{\overset{F}{\underset{\longleftarrow}{\perp \longrightarrow}}} \mathscr{B}$ *be locally small categories, and let* $A \in \mathscr{A}$. *Then the functor*

$$\mathscr{A}(A, G(-))\colon \mathscr{B} \to \mathbf{Set}$$

(that is, the composite $\mathscr{B} \xrightarrow{G} \mathscr{A} \xrightarrow{H^A} \mathbf{Set}$*) is representable.*

Proof   We have

$$\mathscr{A}(A, G(B)) \cong \mathscr{B}(F(A), B)$$

for each $B \in \mathscr{B}$. If we can show that this isomorphism is natural in B, then we will have proved that $\mathscr{A}(A, G(-))$ is isomorphic to $H^{F(A)}$ and is therefore representable. So, let $B \xrightarrow{q} B'$ be a map in $\mathscr{B}$. We must show that the square

$$
\begin{array}{ccc}
\mathscr{A}(A, G(B)) & \longrightarrow & \mathscr{B}(F(A), B) \\
{\scriptstyle G(q)\circ-}\Big\downarrow & & \Big\downarrow{\scriptstyle q\circ-} \\
\mathscr{A}(A, G(B')) & \longrightarrow & \mathscr{B}(F(A), B')
\end{array}
$$

commutes, where the horizontal arrows are the bijections provided by the adjunction. For $f\colon A \to G(B)$, we have

$$
\begin{array}{ccc}
f & \longmapsto & \bar{f} \\
\Big\downarrow & & \Big\downarrow \\
G(q) \circ f & \longmapsto & \overline{G(q) \circ f,} \quad\overset{q \circ \bar{f}}{}
\end{array}
$$

so we must prove that $q \circ \bar{f} = \overline{G(q) \circ f}$. This follows immediately from the naturality condition (2.2) in the definition of adjunction (with $g = \bar{f}$). $\square$

You would not expect a randomly-chosen functor into **Set** to be representable. In some sense, rather few functors are. However, forgetful functors do tend to be representable:

Proposition 4.1.11 *Any set-valued functor with a left adjoint is representable.*

Proof Let $G: \mathscr{A} \to$ **Set** be a functor with a left adjoint F. Write 1 for the one-point set. Then

$$G(A) \cong \mathbf{Set}(1, G(A))$$

naturally in $A \in \mathscr{A}$ (by Example 4.1.4), that is, $G \cong \mathbf{Set}(1, G(-))$. So by Lemma 4.1.10, G is representable; indeed, $G \cong H^{F(1)}$. $\square$

Example 4.1.12 Several of the examples of representables mentioned above arise as in Proposition 4.1.11. For instance, $U:$ **Top** $\to$ **Set** has a left adjoint D (Example 2.1.5), and $D(1) \cong 1$, so we recover the result that $U \cong H^1$. Similarly, Exercise 3.2.16 asked you to construct a left adjoint D to the objects functor ob : **Cat** $\to$ **Set**. This functor D satisfies $D(1) \cong 1$, proving again that ob $\cong H^1$.

Example 4.1.13 The forgetful functor $U:$ **Vect**$_k \to$ **Set** is representable, since it has a left adjoint. Indeed, if F denotes the left adjoint then $F(1)$ is the 1-dimensional vector space k, so $U \cong H^k$. This is also easy to see directly: a map from k to a vector space V is uniquely determined by the image of 1, which can be any element of V; hence **Vect**$_k(k, V) \cong U(V)$ naturally in V.

Example 4.1.14 Examples 2.1.3 began with the declaration that forgetful functors between categories of algebraic structures usually have left adjoints. Take the category **CRing** of commutative rings and the forgetful functor $U:$ **CRing** $\to$ **Set**. This general principle suggests that U has a left adjoint, and Proposition 4.1.11 then tells us that U is representable.

Let us see how this works explicitly. Given a set S, let $\mathbb{Z}[S]$ be the ring of polynomials over $\mathbb{Z}$ in commuting variables x_s ($s \in S$). (This was called $F(S)$ in Example 1.2.4(b).) Then $S \mapsto \mathbb{Z}[S]$ defines a functor **Set** $\to$ **CRing**, and this is left adjoint to U. Hence $U \cong H^{\mathbb{Z}[x]}$. Again, this can be verified directly: for any ring R, the maps $\mathbb{Z}[x] \to R$ correspond one-to-one with the elements of R (Exercises 0.13 and 4.1.29).

We have defined, for each object A of our category $\mathscr{A}$, a functor $H^A \in [\mathscr{A}, \mathbf{Set}]$. This describes how A sees the world. As A varies, the view varies. On the other hand, it is always the same world being seen, so the different views from different objects are somehow related. (Compare aerial photos taken from a moving aeroplane, which agree well enough on their overlaps that they can be patched together to make one big picture.) So the family $(H^A)_{A \in \mathscr{A}}$ of 'views' has some consistency to it. What this means is that whenever there is a map between objects A and A', there is also a map between H^A and $H^{A'}$.

Precisely, a map $A' \xrightarrow{f} A$ induces a natural transformation

$$\mathscr{A} \underset{H^{A'}}{\overset{H^A}{\rightrightarrows}} \Downarrow H^f \quad \textbf{Set,}$$

whose B-component (for $B \in \mathscr{A}$) is the function

$$H^A(B) = \mathscr{A}(A, B) \quad \to \quad H^{A'}(B) = \mathscr{A}(A', B)$$
$$p \quad \mapsto \quad p \circ f.$$

Again, H^f goes by a variety of other names: $\mathscr{A}(f, -)$, f^*, and $- \circ f$.

Note the reversal of direction! Each functor H^A is covariant, but they come together to form a *contravariant* functor, as in the following definition.

Definition 4.1.15 Let $\mathscr{A}$ be a locally small category. The functor

$$H^\bullet : \mathscr{A}^{\mathrm{op}} \to [\mathscr{A}, \textbf{Set}]$$

is defined on objects A by $H^\bullet(A) = H^A$ and on maps f by $H^\bullet(f) = H^f$.

The symbol $\bullet$ is another type of blank, like $-$.

All of the definitions presented so far in this chapter can be dualized. At the formal level, this is trivial: reverse all the arrows, so that every $\mathscr{A}$ becomes an $\mathscr{A}^{\mathrm{op}}$ and vice versa. But in our usual examples, the flavour is different. We are no longer asking what objects *see*, but how they are *seen*.

Let us first dualize Definition 4.1.1.

Definition 4.1.16 Let $\mathscr{A}$ be a locally small category and $A \in \mathscr{A}$. We define a functor

$$H_A = \mathscr{A}(-, A) : \mathscr{A}^{\mathrm{op}} \to \textbf{Set}$$

as follows:

- for objects $B \in \mathscr{A}$, put $H_A(B) = \mathscr{A}(B, A)$;
- for maps $B' \xrightarrow{g} B$ in $\mathscr{A}$, define

$$H_A(g) = \mathscr{A}(g, A) = g^* = - \circ g : \mathscr{A}(B, A) \to \mathscr{A}(B', A)$$

by

$$p \mapsto p \circ g$$

for all $p : B \to A$.

If you know about dual vector spaces, this construction will seem familiar. In particular, you will not be surprised that a map $B' \to B$ induces a map in the opposite direction, $H_A(B) \to H_A(B')$.

We now define representability for *contravariant* set-valued functors. Strictly speaking, this is unnecessary, as a contravariant functor on $\mathscr{A}$ is a covariant functor on $\mathscr{A}^{\mathrm{op}}$, and we already know what it means for a covariant set-valued functor to be representable. But it is useful to have a direct definition.

Definition 4.1.17 Let $\mathscr{A}$ be a locally small category. A functor $X \colon \mathscr{A}^{\mathrm{op}} \to$ **Set** is **representable** if $X \cong H_A$ for some $A \in \mathscr{A}$. A **representation** of X is a choice of an object $A \in \mathscr{A}$ and an isomorphism between H_A and X.

Example 4.1.18 There is a functor

$$\mathscr{P} : \mathbf{Set}^{\mathrm{op}} \to \mathbf{Set}$$

sending each set B to its power set $\mathscr{P}(B)$, and defined on maps $g \colon B' \to B$ by $(\mathscr{P}(g))(U) = g^{-1}U$ for all $U \in \mathscr{P}(B)$. (Here $g^{-1}U$ denotes the inverse image or preimage of U under g, defined by $g^{-1}U = \{x' \in B' \mid g(x') \in U\}$.) As we saw in Section 3.1, a subset amounts to a map into the two-point set 2. Precisely put, $\mathscr{P} \cong H_2$.

Example 4.1.19 Similarly, there is a functor

$$\mathscr{O} : \mathbf{Top}^{\mathrm{op}} \to \mathbf{Set}$$

defined on objects B by taking $\mathscr{O}(B)$ to be the set of open subsets of B. If S denotes the two-point topological space in which exactly one of the two singleton subsets is open, then continuous maps from a space B into S correspond naturally to open subsets of B (Exercise 4.1.30). Hence $\mathscr{O} \cong H_S$, and $\mathscr{O}$ is representable.

Example 4.1.20 In Example 1.2.11, we defined a functor $C \colon \mathbf{Top}^{\mathrm{op}} \to \mathbf{Ring}$, assigning to each space the ring of continuous real-valued functions on it. The composite functor

$$\mathbf{Top}^{\mathrm{op}} \xrightarrow{\ C\ } \mathbf{Ring} \xrightarrow{\ U\ } \mathbf{Set}$$

is representable, since by definition, $U(C(X)) = \mathbf{Top}(X, \mathbb{R})$ for topological spaces X.

Previously, we assembled the covariant representables $(H^A)_{A \in \mathscr{A}}$ into one big functor $H^\bullet$. We now do the same for the contravariant representables $(H_A)_{A \in \mathscr{A}}$.

Any map $A \xrightarrow{f} A'$ in $\mathscr{A}$ induces a natural transformation

$$\mathscr{A}^{\mathrm{op}} \underset{H_{A'}}{\overset{H_A}{\Downarrow H_f}} \mathbf{Set}$$

(also called $\mathscr{A}(-, f)$, f_* or $f \circ -$), whose component at an object $B \in \mathscr{A}$ is

$$H_A(B) = \mathscr{A}(B, A) \quad \to \quad H_{A'}(B) = \mathscr{A}(B, A')$$
$$p \quad \mapsto \quad f \circ p.$$

Definition 4.1.21 Let $\mathscr{A}$ be a locally small category. The **Yoneda embedding** of $\mathscr{A}$ is the functor

$$H_{\bullet} \colon \mathscr{A} \to [\mathscr{A}^{\mathrm{op}}, \mathbf{Set}]$$

defined on objects A by $H_{\bullet}(A) = H_A$ and on maps f by $H_{\bullet}(f) = H_f$.

Here is a summary of the definitions so far.

For each $A \in \mathscr{A}$, we have a functor $\qquad \mathscr{A} \xrightarrow{H^A} \mathbf{Set}$.

Putting them all together gives a functor $\quad \mathscr{A}^{\mathrm{op}} \xrightarrow{H^{\bullet}} [\mathscr{A}, \mathbf{Set}]$.

For each $A \in \mathscr{A}$, we have a functor $\qquad \mathscr{A}^{\mathrm{op}} \xrightarrow{H_A} \mathbf{Set}$.

Putting them all together gives a functor $\quad \mathscr{A} \xrightarrow{H_{\bullet}} [\mathscr{A}^{\mathrm{op}}, \mathbf{Set}]$.

The second pair of functors is the dual of the first. Both involve contravariance; it cannot be avoided.

In the theory of representable functors, it does not make much difference whether we work with the first or the second pair. Any theorem that we prove about one dualizes to give a theorem about the other. We choose to work with the second pair, the H_As and $H_{\bullet}$. In a sense to be explained, $H_{\bullet}$ 'embeds' $\mathscr{A}$ into $[\mathscr{A}^{\mathrm{op}}, \mathbf{Set}]$. This can be useful, because the category $[\mathscr{A}^{\mathrm{op}}, \mathbf{Set}]$ has some good properties that $\mathscr{A}$ might not have.

Exercise 4.1.27 asks you to prove that $H_{\bullet}$ is injective on isomorphism classes of objects. It is strongly recommended that you do it before reading on, as it encapsulates the key ideas of the rest of this chapter.

There is one more functor to define. It unifies the first and second pairs of functors shown above.

Definition 4.1.22 Let $\mathscr{A}$ be a locally small category. The functor

$$\mathrm{Hom}_{\mathscr{A}} \colon \mathscr{A}^{\mathrm{op}} \times \mathscr{A} \to \mathbf{Set}$$

is defined by

$$
\begin{array}{ccc}
(A, B) & \mapsto & \mathscr{A}(A, B) \\
{\scriptstyle f}\uparrow\downarrow{\scriptstyle g} & \mapsto & \downarrow{\scriptstyle g\circ-\circ f} \\
(A', B') & \mapsto & \mathscr{A}(A', B').
\end{array}
$$

In other words, $\mathrm{Hom}_{\mathscr{A}}(A, B) = \mathscr{A}(A, B)$ and $(\mathrm{Hom}_{\mathscr{A}}(f, g))(p) = g \circ p \circ f$, whenever $A' \xrightarrow{f} A \xrightarrow{p} B \xrightarrow{g} B'$.

Remarks 4.1.23 (a) The existence of the functor $\mathrm{Hom}_{\mathscr{A}}$ is something like the fact that for a metric space (X, d), the metric is itself a continuous map $d \colon X \times X \to \mathbb{R}$. (If we take two points and move each one slightly, the distance between them changes only slightly.)

(b) In terms of Exercise 1.2.25, $\mathrm{Hom}_{\mathscr{A}}$ is the functor $\mathscr{A}^{\mathrm{op}} \times \mathscr{A} \to \mathbf{Set}$ corresponding to the families of functors $(H^A)_{A \in \mathscr{A}}$ and $(H_B)_{B \in \mathscr{A}}$.

(c) In Example 2.1.6, we saw that for any set B, there is an adjunction $(- \times B) \dashv (-)^B$ of functors $\mathbf{Set} \to \mathbf{Set}$. Similarly, for any category $\mathscr{B}$, there is an adjunction $(- \times \mathscr{B}) \dashv [\mathscr{B}, -]$ of functors $\mathbf{CAT} \to \mathbf{CAT}$; in other words, there is a canonical bijection

$$
\mathbf{CAT}(\mathscr{A} \times \mathscr{B}, \mathscr{C}) \cong \mathbf{CAT}(\mathscr{A}, [\mathscr{B}, \mathscr{C}])
$$

for $\mathscr{A}, \mathscr{B}, \mathscr{C} \in \mathbf{CAT}$. Under this bijection, the functors

$$
\mathrm{Hom}_{\mathscr{A}} \colon \mathscr{A}^{\mathrm{op}} \times \mathscr{A} \to \mathbf{Set}, \qquad H^{\bullet} \colon \mathscr{A}^{\mathrm{op}} \to [\mathscr{A}, \mathbf{Set}]
$$

correspond to one another. Thus, $\mathrm{Hom}_{\mathscr{A}}$ carries the same information as $H^{\bullet}$ (or $H_{\bullet}$), presented slightly differently.

Remark 4.1.24 We can now explain the naturality in the definition of adjunction (Definition 2.1.1). Take categories and functors $\mathscr{A} \underset{G}{\overset{F}{\rightleftarrows}} \mathscr{B}$. They give rise to functors

$$
\begin{array}{ccc}
\mathscr{A}^{\mathrm{op}} \times \mathscr{B} & \xrightarrow{\ 1 \times G\ } & \mathscr{A}^{\mathrm{op}} \times \mathscr{A} \\
{\scriptstyle F^{\mathrm{op}} \times 1}\downarrow & & \downarrow{\scriptstyle \mathrm{Hom}_{\mathscr{A}}} \\
\mathscr{B}^{\mathrm{op}} \times \mathscr{B} & \xrightarrow[\ \mathrm{Hom}_{\mathscr{B}}\ ]{} & \mathbf{Set}.
\end{array}
$$

The composite functor $\llcorner\!\rightarrow$ sends (A, B) to $\mathscr{B}(F(A), B)$; it can be written as $\mathscr{B}(F(-), -)$. The composite $\rightarrow\!\lrcorner$ sends (A, B) to $\mathscr{A}(A, G(B))$. Exercise 4.1.32 asks you to show that these two functors

$$
\mathscr{B}(F(-), -), \ \mathscr{A}(-, G(-)) \colon \mathscr{A}^{\mathrm{op}} \times \mathscr{B} \to \mathbf{Set}
$$

are naturally isomorphic if and only if F and G are adjoint. This justifies the claim in Remark 2.1.2(a): the naturality requirements (2.2) and (2.3) in the definition of adjunction simply assert that two particular functors are naturally isomorphic.

Objects of an arbitrary category do not have elements in any obvious sense. However, *sets* certainly have elements, and we have observed that an element of a set A is the same thing as a map $1 \to A$. This inspires the following definition.

Definition 4.1.25 Let A be an object of a category. A **generalized element** of A is a map with codomain A. A map $S \to A$ is a generalized element of A of **shape** S.

'Generalized element' is nothing more than a synonym of 'map', but sometimes it is useful to think of maps as generalized elements.

For example, when A is a set, a generalized element of A of shape 1 is an ordinary element of A, and a generalized element of A of shape $\mathbb{N}$ is a sequence in A. In the category of topological spaces, the generalized elements of shape 1 (the one-point space) are the points, and the generalized elements of shape S^1 (the circle) are, by definition, loops. As this suggests, in categories of geometric objects, we might equally well say 'figures of shape S'.

In algebra, we are often interested in solutions to equations such as $x^2 + y^2 = 1$. Perhaps we begin by being particularly interested in solutions in $\mathbb{Q}$, but then realize that in order to study rational solutions, it will be helpful to study solutions in other rings first. (This is often a fruitful strategy.) Given a ring A, a pair $(a, b) \in A \times A$ satisfying $a^2 + b^2 = 1$ amounts to a homomorphism of rings

$$\mathbb{Z}[x, y]/(x^2 + y^2 - 1) \to A.$$

Thus, the solutions to our equation (in any ring) can be seen as the generalized elements of shape $\mathbb{Z}[x, y]/(x^2 + y^2 - 1)$.

For an object S of a category $\mathscr{A}$, the functor

$$H^S : \mathscr{A} \to \mathbf{Set}$$

sends an object to its set of generalized elements of shape S. The functoriality tells us that any map $A \to B$ in $\mathscr{A}$ transforms S-elements of A into S-elements of B. For example, taking $\mathscr{A} = \mathbf{Top}$ and $S = S^1$, any continuous map $A \to B$ transforms loops in A into loops in B.

Exercises

4.1.26 Find three examples of representable functors not mentioned above.

4.1.27 Let $\mathscr{A}$ be a locally small category, and let $A, A' \in \mathscr{A}$ with $H_A \cong H_{A'}$. Prove directly that $A \cong A'$.

4.1.28 Let p be a prime number. Show that the functor $U_p\colon \mathbf{Grp} \to \mathbf{Set}$ defined in Example 4.1.5 is isomorphic to $\mathbf{Grp}(\mathbb{Z}/p\mathbb{Z}, -)$. (To check that there is an isomorphism of functors – that is, a *natural* isomorphism – you will first need to define U_p on maps. There is only one sensible way to do this.)

4.1.29 Using the result of Exercise 0.13(a), prove that the forgetful functor $\mathbf{CRing} \to \mathbf{Set}$ is isomorphic to $\mathbf{CRing}(\mathbb{Z}[x], -)$, as in Example 4.1.14.

4.1.30 The **Sierpiński space** is the two-point topological space S in which one of the singleton subsets is open but the other is not. Prove that for any topological space X, there is a canonical bijection between the open subsets of X and the continuous maps $X \to S$. Use this to show that the functor $\mathcal{O}\colon \mathbf{Top}^{\mathrm{op}} \to \mathbf{Set}$ of Example 4.1.19 is represented by S.

4.1.31 Let $M\colon \mathbf{Cat} \to \mathbf{Set}$ be the functor that sends a small category $\mathscr{A}$ to the set of all maps in $\mathscr{A}$. Prove that M is representable.

4.1.32 Take locally small categories $\mathscr{A}$ and $\mathscr{B}$, and functors $\mathscr{A} \underset{G}{\overset{F}{\rightleftarrows}} \mathscr{B}$. Show that F is left adjoint to G if and only if the two functors

$$\mathscr{B}(F(-), -), \quad \mathscr{A}(-, G(-))\colon \mathscr{A}^{\mathrm{op}} \times \mathscr{B} \to \mathbf{Set}$$

of Remark 4.1.24 are naturally isomorphic. (Hint: this is made easier by using either Exercise 1.3.29 or Exercise 2.1.14.)

4.2 The Yoneda lemma

What do representables see?

Recall from Definition 1.2.15 that functors $\mathscr{A}^{\mathrm{op}} \to \mathbf{Set}$ are sometimes called 'presheaves' on $\mathscr{A}$. So for each $A \in \mathscr{A}$ we have a representable presheaf H_A, and we are asking how the rest of the presheaf category $[\mathscr{A}^{\mathrm{op}}, \mathbf{Set}]$ looks from the viewpoint of H_A. In other words, if X is another presheaf, what are the maps $H_A \to X$?

Newcomers to category theory commonly find that the material presented in this section is where they first get stuck. Typically, the core of the difficulty is in understanding the question just asked. Let us ask it again.

We start by fixing a locally small category $\mathscr{A}$. We then take an object $A \in \mathscr{A}$ and a functor $X\colon \mathscr{A}^{\mathrm{op}} \to \mathbf{Set}$. The object A gives rise to another functor

$H_A = \mathscr{A}(-, A) \colon \mathscr{A}^{\mathrm{op}} \to \mathbf{Set}$. The question is: what are the maps $H_A \to X$? Since H_A and X are both objects of the presheaf category $[\mathscr{A}^{\mathrm{op}}, \mathbf{Set}]$, the 'maps' concerned are maps in $[\mathscr{A}^{\mathrm{op}}, \mathbf{Set}]$. So, we are asking what natural transformations

$$\mathscr{A}^{\mathrm{op}} \underset{X}{\overset{H_A}{\rightrightarrows}} \Downarrow \mathbf{Set} \tag{4.1}$$

there are. The set of such natural transformations is called

$$[\mathscr{A}^{\mathrm{op}}, \mathbf{Set}](H_A, X).$$

(This is a special case of the notation $\mathscr{B}(B, B')$ for the set of maps $B \to B'$ in a category $\mathscr{B}$. Here, $\mathscr{B} = [\mathscr{A}^{\mathrm{op}}, \mathbf{Set}]$, $B = H_A$, and $B' = X$.) We want to know what this set is.

There is an informal principle of general category theory that allows us to guess the answer. Look back at Remarks 1.1.2(b), 1.2.2(a) and 1.3.2(a) on the definitions of category, functor and natural transformation. Each remark is of the form 'from input of one type, it is possible to construct exactly one output of another type'. For example, in Remark 1.1.2(b), the input is a sequence of maps $A_0 \xrightarrow{f_1} \cdots \xrightarrow{f_n} A_n$, the output is a map $A_0 \to A_n$, and the statement is that no matter what we do with the input data $f_1, \ldots, f_n$, there is only one map $A_0 \to A_n$ that we can construct.

Let us apply this principle to our question. We have just seen how, given as input an object $A \in \mathscr{A}$ and a presheaf X on $\mathscr{A}$, we can construct a set, namely, $[\mathscr{A}^{\mathrm{op}}, \mathbf{Set}](H_A, X)$. Are there any other ways to construct a set from the same input data (A, X)? Yes: simply take the set $X(A)$! The informal principle suggests that these two sets are the same:

$$[\mathscr{A}^{\mathrm{op}}, \mathbf{Set}](H_A, X) \cong X(A) \tag{4.2}$$

for all $A \in \mathscr{A}$ and $X \in [\mathscr{A}^{\mathrm{op}}, \mathbf{Set}]$. This turns out to be true; and that is the Yoneda lemma.

Informally, then, the Yoneda lemma says that for any $A \in \mathscr{A}$ and presheaf X on $\mathscr{A}$:

A natural transformation $H_A \to X$ is an element of $X(A)$.

Here is the formal statement. The proof follows shortly.

Theorem 4.2.1 (Yoneda) *Let $\mathscr{A}$ be a locally small category. Then*

$$[\mathscr{A}^{\mathrm{op}}, \mathbf{Set}](H_A, X) \cong X(A) \tag{4.3}$$

naturally in $A \in \mathscr{A}$ and $X \in [\mathscr{A}^{\mathrm{op}}, \mathbf{Set}]$.

This is exactly what was stated in (4.2), except that the word 'naturally' has appeared. Recall from Definition 1.3.12 that for functors $F, G: \mathscr{C} \to \mathscr{D}$, the phrase '$F(C) \cong G(C)$ naturally in C' means that there is a natural isomorphism $F \cong G$. So the use of this phrase in the Yoneda lemma suggests that each side of (4.3) is functorial in both A and X. This means, for instance, that a map $X \to X'$ must induce a map

$$[\mathscr{A}^{\mathrm{op}}, \mathbf{Set}](H_A, X) \to [\mathscr{A}^{\mathrm{op}}, \mathbf{Set}](H_A, X'),$$

and that not only does the isomorphism (4.3) hold for *every* A and X, but also, the isomorphisms can be chosen in a way that is compatible with these induced maps. Precisely, the Yoneda lemma states that the composite functor

$$\mathscr{A}^{\mathrm{op}} \times [\mathscr{A}^{\mathrm{op}}, \mathbf{Set}] \xrightarrow{H_{\bullet}^{\mathrm{op}} \times 1} [\mathscr{A}^{\mathrm{op}}, \mathbf{Set}]^{\mathrm{op}} \times [\mathscr{A}^{\mathrm{op}}, \mathbf{Set}] \xrightarrow{\mathrm{Hom}_{[\mathscr{A}^{\mathrm{op}}, \mathbf{Set}]}} \mathbf{Set}$$
$$(A, X) \quad\longmapsto\quad (H_A, X) \quad\longmapsto\quad [\mathscr{A}^{\mathrm{op}}, \mathbf{Set}](H_A, X)$$

is naturally isomorphic to the evaluation functor

$$\mathscr{A}^{\mathrm{op}} \times [\mathscr{A}^{\mathrm{op}}, \mathbf{Set}] \quad\to\quad \mathbf{Set}$$
$$(A, X) \quad\mapsto\quad X(A).$$

If the Yoneda lemma were false then the world would look much more complex. For take a presheaf $X: \mathscr{A}^{\mathrm{op}} \to \mathbf{Set}$, and define a new presheaf X' by

$$X' = [\mathscr{A}^{\mathrm{op}}, \mathbf{Set}](H_{\bullet}, X): \mathscr{A}^{\mathrm{op}} \to \mathbf{Set},$$

that is, $X'(A) = [\mathscr{A}^{\mathrm{op}}, \mathbf{Set}](H_A, X)$ for all $A \in \mathscr{A}$. Yoneda tells us that $X'(A) \cong X(A)$ naturally in A; in other words, $X' \cong X$. If Yoneda were false then starting from a single presheaf X, we could build an infinite sequence $X, X', X'', \ldots$ of new presheaves, potentially all different. But in reality, the situation is very simple: they are all the same.

The proof of the Yoneda lemma is the longest proof so far. Nevertheless, there is essentially only one way to proceed at each stage. If you suspect that you are one of those newcomers to category theory for whom the Yoneda lemma presents the first serious challenge, an excellent exercise is to work out the proof before reading it. No ingenuity is required, only an understanding of all the terms in the statement.

Proof of the Yoneda lemma We have to define, for each A and X, a bijection between the sets $[\mathscr{A}^{\mathrm{op}}, \mathbf{Set}](H_A, X)$ and $X(A)$. We then have to show that our bijection is natural in A and X.

First, fix $A \in \mathscr{A}$ and $X \in [\mathscr{A}^{op}, \mathbf{Set}]$. We define functions

$$[\mathscr{A}^{op}, \mathbf{Set}](H_A, X) \underset{(\tilde{\ })}{\overset{(\hat{\ })}{\rightleftarrows}} X(A) \qquad (4.4)$$

and show that they are mutually inverse. So we have to do four things: define the function $(\hat{\ })$, define the function $(\tilde{\ })$, show that $(\hat{\tilde{\ }})$ is the identity, and show that $(\tilde{\hat{\ }})$ is the identity.

- Given $\alpha\colon H_A \to X$, define $\hat{\alpha} \in X(A)$ by $\hat{\alpha} = \alpha_A(1_A)$. (How else could we possibly define it?)
- Let $x \in X(A)$. We have to define a natural transformation $\tilde{x}\colon H_A \to X$. That is, we have to define for each $B \in \mathscr{A}$ a function

$$\tilde{x}_B\colon H_A(B) = \mathscr{A}(B, A) \to X(B)$$

and show that the family $\tilde{x} = (\tilde{x}_B)_{B \in \mathscr{A}}$ satisfies naturality.

Given $B \in \mathscr{A}$ and $f \in \mathscr{A}(B, A)$, define

$$\tilde{x}_B(f) = (X(f))(x) \in X(B).$$

(How else could we possibly define it?) This makes sense, since $X(f)$ is a map $X(A) \to X(B)$. To prove naturality, we must show that for any map $B' \xrightarrow{g} B$ in $\mathscr{A}$, the square

$$
\begin{array}{ccc}
\mathscr{A}(B, A) & \xrightarrow{\ H_A(g)\, =\, -\circ g\ } & \mathscr{A}(B', A) \\
{\scriptstyle \tilde{x}_B}\downarrow & & \downarrow{\scriptstyle \tilde{x}_{B'}} \\
X(B) & \xrightarrow[\ X(g)\ ]{} & X(A)
\end{array}
$$

commutes. To reduce clutter, let us write $X(g)$ as Xg, and so on. Now for all $f \in \mathscr{A}(B, A)$, we have

$$
\begin{array}{ccc}
f & \longmapsto & f \circ g \\
\downarrow & & \downarrow \\
& & (X(f \circ g))(x) \\
(Xf)(x) & \longmapsto & (Xg)((Xf)(x)),
\end{array}
$$

and $X(f \circ g) = (Xg) \circ (Xf)$ by functoriality, so the square does commute.

- Given $x \in X(A)$, we have to show that $\hat{\tilde{x}} = x$, and indeed,

$$\hat{\tilde{x}} = \tilde{x}_A(1_A) = (X1_A)(x) = 1_{X(A)}(x) = x.$$

- Given $\alpha\colon H_A \to X$, we have to show that $\tilde{\hat{\alpha}} = \alpha$. Two natural transformations are equal if and only if all their components are equal; so, we have to show that $\left(\tilde{\hat{\alpha}}\right)_B = \alpha_B$ for all $B \in \mathscr{A}$. Each side of this equation is a function from $H_A(B) = \mathscr{A}(B, A)$ to $X(B)$, and two functions are equal if and only if they take equal values at every element of the domain; so, we have to show that

$$\left(\tilde{\hat{\alpha}}\right)_B(f) = \alpha_B(f)$$

for all $B \in \mathscr{A}$ and $f\colon B \to A$ in $\mathscr{A}$. The left-hand side is by definition

$$\left(\tilde{\hat{\alpha}}\right)_B(f) = (Xf)(\hat{\alpha}) = (Xf)(\alpha_A(1_A)),$$

so it remains to prove that

$$(Xf)(\alpha_A(1_A)) = \alpha_B(f). \tag{4.5}$$

By naturality of α (the only tool at our disposal), the square

$$
\begin{array}{ccc}
\mathscr{A}(A,A) & \xrightarrow{\;H_A(f)\,=\,-\circ f\;} & \mathscr{A}(B,A) \\
{\scriptstyle \alpha_A}\downarrow & & \downarrow{\scriptstyle \alpha_B} \\
X(A) & \xrightarrow[\;Xf\;]{} & X(B)
\end{array}
$$

commutes, which when taken at $1_A \in \mathscr{A}(A,A)$ gives equation (4.5).

(The proof is not over yet, but it is worth pausing to consider the significance of the fact that $\tilde{\hat{\alpha}} = \alpha$. Since $\hat{\alpha}$ is the value of α at 1_A, this implies:

A natural transformation $H_A \to X$ is determined by its value at 1_A.

Just *how* a natural transformation $H_A \to X$ is determined by its value at 1_A is described in equation (4.5).)

This establishes the bijection (4.4) for each $A \in \mathscr{A}$ and $X \in [\mathscr{A}^{\mathrm{op}}, \mathbf{Set}]$. We now show that the bijection is natural in A and X.

We employ two mildly labour-saving devices. First, in principle we have to prove naturality of both $(\,\hat{}\,)$ and $(\,\tilde{}\,)$, but by Lemma 1.3.11, it is enough to prove naturality of just one of them. We prove naturality of $(\,\hat{}\,)$. Second, by Exercise 1.3.29, $(\,\hat{}\,)$ is natural in the pair (A, X) if and only if it is natural in A for each fixed X and natural in X for each fixed A. So, it remains to check these two types of naturality.

Naturality in A states that for each $X \in [\mathscr{A}^{\mathrm{op}}, \mathbf{Set}]$ and $B \xrightarrow{f} A$ in $\mathscr{A}$, the square

$$[\mathscr{A}^{\mathrm{op}}, \mathbf{Set}](H_A, X) \xrightarrow{\;-\circ H_f\;} [\mathscr{A}^{\mathrm{op}}, \mathbf{Set}](H_B, X)$$

$$(\hat{\ })\Big\downarrow \qquad\qquad\qquad\qquad \Big\downarrow(\hat{\ })$$

$$X(A) \xrightarrow[\;Xf\;]{} X(B)$$

commutes. For $\alpha \colon H_A \to X$, we have

$$\alpha \longmapsto \alpha \circ H_f$$
$$\Big\downarrow \qquad\qquad\qquad\qquad \Big\downarrow$$
$$\qquad\qquad\qquad (\alpha \circ H_f)_B(1_B)$$
$$\alpha_A(1_A) \longmapsto (Xf)(\alpha_A(1_A)),$$

so we have to show that $(\alpha \circ H_f)_B(1_B) = (Xf)(\alpha_A(1_A))$. Indeed,

$$(\alpha \circ H_f)_B(1_B) = \alpha_B((H_f)_B(1_B))$$
$$= \alpha_B(f \circ 1_B) = \alpha_B(f)$$
$$= (Xf)(\alpha_A(1_A)),$$

where the first step is by definition of composition in $[\mathscr{A}^{\mathrm{op}}, \mathbf{Set}]$, the second is by definition of H_f, and the last is by equation (4.5).

Naturality in X states that for each $A \in \mathscr{A}$ and map

$$\mathscr{A}^{\mathrm{op}} \underset{X'}{\overset{X}{\Rightarrow\theta}} \mathbf{Set}$$

in $[\mathscr{A}^{\mathrm{op}}, \mathbf{Set}]$, the square

$$[\mathscr{A}^{\mathrm{op}}, \mathbf{Set}](H_A, X) \xrightarrow{\;\theta\circ-\;} [\mathscr{A}^{\mathrm{op}}, \mathbf{Set}](H_A, X')$$

$$(\hat{\ })\Big\downarrow \qquad\qquad\qquad\qquad \Big\downarrow(\hat{\ })$$

$$X(A) \xrightarrow[\;\theta_A\;]{} X'(A)$$

commutes. For $\alpha \colon H_A \to X$, we have

$$\alpha \longmapsto \theta \circ \alpha$$
$$\Big\downarrow \qquad\qquad\qquad\qquad \Big\downarrow$$
$$\qquad\qquad\qquad (\theta \circ \alpha)_A(1_A)$$
$$\alpha_A(1_A) \longmapsto \theta_A(\alpha_A(1_A)),$$

and $(\theta \circ \alpha)_A = \theta_A \circ \alpha_A$ by definition of composition in $[\mathscr{A}^{\mathrm{op}}, \mathbf{Set}]$, so the square does commute. This completes the proof. $\qquad\qquad\square$

Exercises

4.2.2 State the dual of the Yoneda lemma.

4.2.3 One way to understand the Yoneda lemma is to examine some special cases. Here we consider one-object categories.

Let M be a monoid. The underlying set of M can be given a right M-action by multiplication: $x \cdot m = xm$ for all $x, m \in M$. This M-set is called the **right regular representation** of M. Let us write it as $\underline{M}$.

(a) When M is regarded as a one-object category, functors $M^{\mathrm{op}} \to \mathbf{Set}$ correspond to right M-sets (Example 1.2.14). Show that the M-set corresponding to the unique representable functor $M^{\mathrm{op}} \to \mathbf{Set}$ is the right regular representation.

(b) Now let X be any right M-set. Show that for each $x \in X$, there is a unique map $\alpha \colon \underline{M} \to X$ of right M-sets such that $\alpha(1) = x$. Deduce that there is a bijection between {maps $\underline{M} \to X$ of right M-sets} and X.

(c) Deduce the Yoneda lemma for one-object categories.

4.3 Consequences of the Yoneda lemma

The Yoneda lemma is fundamental in category theory. Here we look at three important consequences.

Notation 4.3.1 An arrow decorated with a $\sim$, as in $A \xrightarrow{\sim} B$, denotes an isomorphism.

A representation is a universal element

Corollary 4.3.2 *Let $\mathscr{A}$ be a locally small category and $X \colon \mathscr{A}^{\mathrm{op}} \to \mathbf{Set}$. Then a representation of X consists of an object $A \in \mathscr{A}$ together with an element $u \in X(A)$ such that:*

$$\text{for each } B \in \mathscr{A} \text{ and } x \in X(B), \text{ there is a unique map } \bar{x} \colon B \to A \\ \text{such that } (X\bar{x})(u) = x. \tag{4.6}$$

To clarify the statement, first recall that by definition, a representation of X is an object $A \in \mathscr{A}$ together with a natural isomorphism $\alpha \colon H_A \xrightarrow{\sim} X$. Corollary 4.3.2 states that such pairs (A, α) are in natural bijection with pairs (A, u) satisfying condition (4.6).

Pairs (B, x) with $B \in \mathscr{A}$ and $x \in X(B)$ are sometimes called **elements** of the presheaf X. (Indeed, the Yoneda lemma tells us that x amounts to a generalized element of X of shape H_B.) An element u satisfying condition (4.6) is sometimes called a **universal** element of X. So, Corollary 4.3.2 says that a representation of a presheaf X amounts to a universal element of X.

Proof By the Yoneda lemma, we have only to show that for $A \in \mathscr{A}$ and $u \in X(A)$, the natural transformation $\bar{u} \colon H_A \to X$ is an isomorphism if and only if (4.6) holds. (Here we are using the notation introduced in the proof of the Yoneda lemma.) Now, $\bar{u}$ is an isomorphism if and only if for all $B \in \mathscr{A}$, the function

$$\bar{u}_B \colon H_A(B) = \mathscr{A}(B, A) \to X(B)$$

is a bijection, if and only if for all $B \in \mathscr{A}$ and $x \in X(B)$, there is a unique $\bar{x} \in \mathscr{A}(B, A)$ such that $\bar{u}_B(\bar{x}) = x$. But $\bar{u}_B(\bar{x}) = (X\bar{x})(u)$, so this is exactly condition (4.6). $\square$

Our examples will use the dual form, for covariant set-valued functors:

Corollary 4.3.3 *Let $\mathscr{A}$ be a locally small category and $X \colon \mathscr{A} \to$ **Set**. Then a representation of X consists of an object $A \in \mathscr{A}$ together with an element $u \in X(A)$ such that:*

for each $B \in \mathscr{A}$ and $x \in X(B)$, there is a unique map $\bar{x} \colon A \to B$ such that $(X\bar{x})(u) = x$. $\quad$ (4.7)

Proof Follows immediately by duality. $\square$

Example 4.3.4 Fix a set S and consider the functor

$$X = \mathbf{Set}(S, U(-)) \colon \quad \mathbf{Vect}_k \quad \to \quad \mathbf{Set}$$
$$V \quad \mapsto \quad \mathbf{Set}(S, U(V)).$$

Here are two familiar (and true!) statements about X:

(a) there exist a vector space $F(S)$ and an isomorphism

$$\mathbf{Vect}_k(F(S), V) \cong \mathbf{Set}(S, U(V)) \qquad (4.8)$$

natural in $V \in \mathbf{Vect}_k$ (Example 2.1.3(a));

(b) there exist a vector space $F(S)$ and a function $u \colon S \to U(F(S))$ such that:

for each vector space V and function $f \colon S \to U(V)$, there is a unique linear map $\bar{f} \colon F(S) \to V$ such that

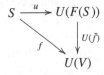

commutes

(as in the introduction to Section 2.3, where u was called by its usual name, η_S).

Each of these two statements says that X is representable. Statement (a) says that there is an isomorphism $X(V) \cong \mathbf{Set}(F(S), V)$ natural in V, that is, an isomorphism $X \cong H^{F(S)}$. So X is representable, by definition of representability. Statement (b) says that $u \in X(F(S))$ satisfies condition (4.7). So X is representable, by Corollary 4.3.3.

You will have noticed that the first way of saying that X is representable is substantially shorter than the second. Indeed, it is clear that if the situation of (b) holds then there is an isomorphism

$$\mathbf{Vect}_k(F(S), V) \xrightarrow{\sim} \mathbf{Set}(S, U(V))$$

natural in V, defined by $g \mapsto U(g) \circ u$. But it looks at first as if (b) says rather more than (a), since it states that the two functors are not only naturally isomorphic, but naturally isomorphic in a rather special way. Corollary 4.3.3 tells us that this is an illusion: all natural isomorphisms (4.8) arise in this way. It is the word 'natural' in (a) that hides the explicit detail.

Example 4.3.5 The same can be said for any other adjunction $\mathscr{A} \underset{G}{\overset{F}{\underset{\perp}{\rightleftarrows}}} \mathscr{B}$.

Fix $A \in \mathscr{A}$ and put

$$X = \mathscr{A}(A, G(-)) \colon \mathscr{B} \to \mathbf{Set}.$$

Then X is representable, and this can be expressed in either of the following ways:

(a) $\mathscr{A}(A, G(B)) \cong \mathscr{B}(F(A), B)$ naturally in B; in other words, $X \cong H^{F(A)}$ (as in Lemma 4.1.10);

(b) the unit map $\eta_A \colon A \to G(F(A))$ is an initial object of the comma category $(A \Rightarrow G)$; that is, $\eta_A \in X(F(A))$ satisfies condition (4.7).

This observation can be developed into an alternative proof of Theorem 2.3.6, the reformulation of adjointness in terms of initial objects.

Example 4.3.6 For any group G and element $x \in G$, there is a unique homomorphism $\phi \colon \mathbb{Z} \to G$ such that $\phi(1) = x$. This means that $1 \in U(\mathbb{Z})$ is a universal element of the forgetful functor $U \colon \mathbf{Grp} \to \mathbf{Set}$; in other words, condition (4.7) holds when $\mathscr{A} = \mathbf{Grp}$, $X = U$, $A = \mathbb{Z}$ and $u = 1$. So $1 \in U(\mathbb{Z})$ gives a representation $H^{\mathbb{Z}} \xrightarrow{\sim} U$ of U.

On the other hand, the same is true with -1 in place of 1. The isomorphisms

$H^Z \xrightarrow{\sim} U$ coming from 1 and -1 are not equal, because Corollary 4.3.3 provides a *one-to-one* correspondence between universal elements and representations.

The Yoneda embedding

Here is a second corollary of the Yoneda lemma.

Corollary 4.3.7 *For any locally small category $\mathscr{A}$, the Yoneda embedding*

$$H_\bullet \colon \mathscr{A} \to [\mathscr{A}^{op}, \mathbf{Set}]$$

is full and faithful.

Informally, this says that for $A, A' \in \mathscr{A}$, a map $H_A \to H_{A'}$ of presheaves is the same thing as a map $A \to A'$ in $\mathscr{A}$.

Proof We have to show that for each $A, A' \in \mathscr{A}$, the function

$$
\begin{array}{ccc}
\mathscr{A}(A, A') & \to & [\mathscr{A}^{op}, \mathbf{Set}](H_A, H_{A'}) \\
f & \mapsto & H_f
\end{array}
\tag{4.9}
$$

is bijective. By the Yoneda lemma (taking 'X' to be $H_{A'}$), the function

$$(\tilde{\ }) \colon H_{A'}(A) \to [\mathscr{A}^{op}, \mathbf{Set}](H_A, H_{A'}) \tag{4.10}$$

is bijective, so it is enough to prove that the functions (4.9) and (4.10) are equal. Thus, given $f \colon A \to A'$, we have to prove that $\tilde{f} = H_f$, or equivalently, $\widehat{H_f} = f$. And indeed,

$$\widehat{H_f} = (H_f)_A(1_A) = f \circ 1_A = f,$$

as required. $\square$

In mathematics at large, the word 'embedding' is used (sometimes informally) to mean a map $A \to B$ that makes A isomorphic to its image in B. For example, an injection of sets $i \colon A \to B$ might be called an embedding, because it provides a bijection between A and the subset iA of B. Similarly, a map $i \colon A \to B$ of topological spaces might be called an embedding if it is a homeomorphism to its image, so that $A \cong iA$. Corollary 1.3.19 tells us that in category theory, a full and faithful functor $\mathscr{A} \to \mathscr{B}$ can reasonably be called an embedding, as it makes $\mathscr{A}$ equivalent to a full subcategory of $\mathscr{B}$.

In the case at hand, the Yoneda embedding $H_\bullet \colon \mathscr{A} \to [\mathscr{A}^{op}, \mathbf{Set}]$ embeds $\mathscr{A}$ into its own presheaf category (Figure 4.1). So, $\mathscr{A}$ is equivalent to the full subcategory of $[\mathscr{A}^{op}, \mathbf{Set}]$ whose objects are the representables.

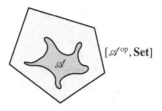

Figure 4.1 A category $\mathscr{A}$ embedded into its presheaf category.

In general, full subcategories are the easiest subcategories to handle. For instance, given objects A and A' of a full subcategory, we can speak unambiguously of the 'maps' from A to A'; it makes no difference whether this is understood to mean maps in the subcategory or maps in the whole category. Similarly, we can speak unambiguously of isomorphism of objects of the subcategory, as in the following lemma.

Lemma 4.3.8 *Let $J\colon \mathscr{A} \to \mathscr{B}$ be a full and faithful functor and $A, A' \in \mathscr{A}$. Then:*

(a) *a map f in $\mathscr{A}$ is an isomorphism if and only if the map $J(f)$ in $\mathscr{B}$ is an isomorphism;*

(b) *for any isomorphism $g\colon J(A) \to J(A')$ in $\mathscr{B}$, there is a unique isomorphism $f\colon A \to A'$ in $\mathscr{A}$ such that $J(f) = g$;*

(c) *the objects A and A' of $\mathscr{A}$ are isomorphic if and only if the objects $J(A)$ and $J(A')$ of $\mathscr{B}$ are isomorphic.*

Proof Exercise 4.3.15. □

Example 4.3.9 In Example 4.3.6, we considered the representations of the forgetful functor $U\colon \mathbf{Grp} \to \mathbf{Set}$, and found two different isomorphisms $H^{\mathbb{Z}} \xrightarrow{\sim} U$. Did we find all of them?

Since $H^{\mathbb{Z}} \cong U$, there are as many isomorphisms $H^{\mathbb{Z}} \xrightarrow{\sim} U$ as there are isomorphisms $H^{\mathbb{Z}} \xrightarrow{\sim} H^{\mathbb{Z}}$. By Corollary 4.3.7 and Lemma 4.3.8(b), there are as many of *these* as there are group isomorphisms $\mathbb{Z} \xrightarrow{\sim} \mathbb{Z}$. There are precisely two such (corresponding to the two generators ± 1 of $\mathbb{Z}$), so we did indeed find all the isomorphisms $H^{\mathbb{Z}} \xrightarrow{\sim} U$. Differently put, there are exactly two universal elements of $U(\mathbb{Z})$.

In Section 6.2, we will see that every presheaf can be built from representables, in very roughly the same way that every positive integer can be built from primes.

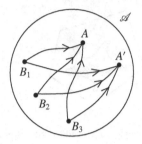

Figure 4.2 If $\mathscr{A}(B, A) \cong \mathscr{A}(B, A')$ naturally in B, then $A \cong A'$.

Isomorphism of representables

In Exercise 4.1.27, you were asked to prove directly that if $H_A \cong H_{A'}$ then $A \cong A'$. The proof contains all the main ideas in the proof of the Yoneda lemma. The result itself can also be deduced from the Yoneda lemma, as follows.

Corollary 4.3.10 *Let $\mathscr{A}$ be a locally small category and $A, A' \in \mathscr{A}$. Then*

$$H_A \cong H_{A'} \iff A \cong A' \iff H^A \cong H^{A'}.$$

Proof By duality, it is enough to prove the first '$\iff$'. This follows from Corollary 4.3.7 and Lemma 4.3.8(c). □

Since functors always preserve isomorphism (Exercise 1.2.21), the force of this statement is that

$$H_A \cong H_{A'} \implies A \cong A'.$$

In other words, if $\mathscr{A}(B, A) \cong \mathscr{A}(B, A')$ naturally in B, then $A \cong A'$. Thinking of $\mathscr{A}(B, A)$ as 'A viewed from B', the corollary tells us that two objects are the same if and only if they look the same from all viewpoints (Figure 4.2). (If it looks like a duck, walks like a duck, and quacks like a duck, then it probably is a duck.)

Example 4.3.11 Consider Corollary 4.3.10 in the case $\mathscr{A} = \mathbf{Grp}$. Take two groups A and A', and suppose someone tells us that A and A' 'look the same from B' (meaning that $H_A(B) \cong H_{A'}(B)$) for all groups B. Then, for instance:

- $H_A(1) \cong H_{A'}(1)$, where 1 is the trivial group. But $H_A(1) = \mathbf{Grp}(1, A)$ is a one-element set, as is $H_{A'}(1)$, no matter what A and A' are. So this tells us nothing at all.
- $H_A(\mathbb{Z}) \cong H_{A'}(\mathbb{Z})$. We know that $H_A(\mathbb{Z})$ is the underlying set of A, and similarly for A'. So A and A' have isomorphic underlying sets. But for all we know so far, they might have entirely different group structures.

- $H_A(\mathbb{Z}/p\mathbb{Z}) \cong H_{A'}(\mathbb{Z}/p\mathbb{Z})$ for every prime p, so by Example 4.1.5, A and A' have the same number of elements of each prime order.

Each of these isomorphisms gives only partial information about the similarity of A and A'. But if we know that $H_A(B) \cong H_{A'}(B)$ for all groups B, and *naturally* in B, then $A \cong A'$.

Example 4.3.12 The category of sets is very unusual in this respect. For any set A, we have

$$A \cong \mathbf{Set}(1, A) = H_A(1),$$

so $H_A(1) \cong H_{A'}(1)$ implies $A \cong A'$. In other words, two objects of **Set** are the same if they look the same from the point of view of the one-element set. This is a familiar feature of sets: the only thing that matters about a set is its elements!

For a general category, Corollary 4.3.10 tells us that two objects are the same if they have the same generalized elements of all shapes. But the category of sets has a special property: if I choose an object and tell you only what its generalized elements of shape 1 are, then you can deduce exactly what my object must be.

Example 4.3.13 Let $G: \mathscr{B} \to \mathscr{A}$ be a functor, and suppose that both F and F' are left adjoint to G. Then for each $A \in \mathscr{A}$, we have

$$\mathscr{B}(F(A), B) \cong \mathscr{A}(A, G(B)) \cong \mathscr{B}(F'(A), B)$$

naturally in $B \in \mathscr{B}$, so $H^{F(A)} \cong H^{F'(A)}$, so $F(A) \cong F'(A)$ by Corollary 4.3.10. In fact, this isomorphism is natural in A, so that $F \cong F'$. This shows that left adjoints are unique, as claimed in Remark 2.1.2(d). Dually, right adjoints are unique. See also Exercise 4.3.18.

Example 4.3.14 Corollary 4.3.10 implies that if a set-valued functor is isomorphic to both H^A and $H^{A'}$ then $A \cong A'$. So the functor *determines* the representing object, if one exists. For instance, take the functor

$$\mathbf{Bilin}(U, V; -)\colon \mathbf{Vect}_k \to \mathbf{Set}$$

of Example 4.1.9. Corollary 4.3.10 implies that up to isomorphism, there is *at most one* vector space T such that

$$\mathbf{Bilin}(U, V; W) \cong \mathbf{Vect}_k(T, W)$$

naturally in W. It can be shown that there does, in fact, exist such a vector space T. Since all such spaces T are isomorphic, it is legitimate to refer to any of them as *the* tensor product of U and V.

Exercises

4.3.15   Prove Lemma 4.3.8.

4.3.16   Let $\mathscr{A}$ be a locally small category. Prove each of the following statements directly (without using the Yoneda lemma).

(a) $H_{\bullet} \colon \mathscr{A} \to [\mathscr{A}^{\mathrm{op}}, \mathbf{Set}]$ is faithful.

(b) $H_{\bullet}$ is full.

(c) Given $A \in \mathscr{A}$ and a presheaf X on $\mathscr{A}$, if $X(A)$ has an element u that is universal in the sense of Corollary 4.3.2, then $X \cong H_A$.

4.3.17   Interpret the theory of Chapter 4 in the case where the category $\mathscr{A}$ is discrete. For example, what do presheaves look like, and which ones are representable? What does the Yoneda lemma tell us? Does its proof become any shorter? What about the corollaries of the Yoneda lemma?

4.3.18   Let $\mathscr{B}$ be a category and $J \colon \mathscr{C} \to \mathscr{D}$ a functor. There is an induced functor

$$J \circ - \colon [\mathscr{B}, \mathscr{C}] \to [\mathscr{B}, \mathscr{D}]$$

defined by composition with J.

(a) Show that if J is full and faithful then so is $J \circ -$.

(b) Deduce that if J is full and faithful and $G, G' \colon \mathscr{B} \to \mathscr{C}$ with $J \circ G \cong J \circ G'$ then $G \cong G'$.

(c) Now deduce that right adjoints are unique: if $F \colon \mathscr{A} \to \mathscr{B}$ and $G, G' \colon \mathscr{B} \to \mathscr{A}$ with $F \dashv G$ and $F \dashv G'$ then $G \cong G'$. (Hint: the Yoneda embedding is full and faithful.)

5

Limits

Limits, and the dual concept, colimits, provide our third approach to the idea of universal property.

Adjointness is about the relationships *between* categories. Representability is a property of *set-valued* functors. Limits are about what goes on *inside* a category.

The concept of limit unifies many familiar constructions in mathematics. Whenever you meet a method for taking some objects and maps in a category and constructing a new object out of them, there is a good chance that you are looking at either a limit or a colimit. For instance, in group theory, we can take a homomorphism between two groups and form its kernel, which is a new group. This construction is an example of a limit in the category of groups. Or, we might take two natural numbers and form their lowest common multiple. This is an example of a colimit in the poset of natural numbers, ordered by divisibility.

5.1 Limits: definition and examples

The definition of limit is very general. We build up to it by first examining some particularly useful types of limit: products, equalizers, and pullbacks.

Products

Let X and Y be sets. The familiar cartesian product $X \times Y$ is characterized by the property that an element of $X \times Y$ is an element of X together with an element of Y. Since elements are just maps from 1, this says that a map $1 \to X \times Y$ amounts to a map $1 \to X$ together with a map $1 \to Y$.

A little thought reveals that the same is true when 1 is replaced throughout

107

by any set A whatsoever. (In other words, a generalized element of $X \times Y$ of shape A amounts to a generalized element of X of shape A together with a generalized element of Y of shape A.) The bijection between

$$\text{maps } A \to X \times Y$$

and

$$\text{pairs of maps } (A \to X, A \to Y)$$

is given by composing with the projection maps

$$X \xleftarrow{p_1} X \times Y \xrightarrow{p_2} Y$$
$$x \leftarrowtail (x, y) \mapsto y.$$

This suggests the following definition.

Definition 5.1.1   Let $\mathscr{A}$ be a category and $X, Y \in \mathscr{A}$. A **product** of X and Y consists of an object P and maps

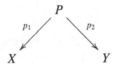

with the property that for all objects and maps

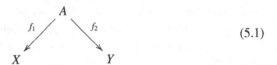   (5.1)

in $\mathscr{A}$, there exists a unique map $\bar{f} \colon A \to P$ such that

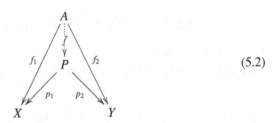   (5.2)

commutes. The maps p_1 and p_2 are called the **projections**.

Remarks 5.1.2   (a) Products do not always exist. For example, if $\mathscr{A}$ is the discrete two-object category

$$\boxed{X \bullet \qquad \bullet Y}$$

then X and Y do not have a product. But when objects X and Y of a category do have a product, it is unique up to isomorphism. (This can be proved directly, much as in Lemma 2.1.8. It also follows from Corollary 6.1.2.) This justifies talking about *the* product of X and Y.

(b) Strictly speaking, the product consists of the object P *together with* the projections p_1 and p_2. But informally, we often refer to P alone as the product of X and Y. We write P as $X \times Y$.

Example 5.1.3 Any two sets X and Y have a product in **Set**. It is the usual cartesian product $X \times Y$, equipped with the usual projection maps p_1 and p_2.

Let us check that this really is a product in the sense of Definition 5.1.1. Take sets and functions as in diagram (5.1). Define $\bar{f} \colon A \to X \times Y$ by $\bar{f}(a) = (f_1(a), f_2(a))$. Then $p_i \circ \bar{f} = f_i$ for $i = 1, 2$; that is, diagram (5.2) commutes with $P = X \times Y$. Moreover, this is the *only* map making diagram (5.2) commute. For suppose that $\hat{f} \colon A \to X \times Y$, in place of $\bar{f}$, also makes (5.2) commute. Let $a \in A$, and write $\hat{f}(a)$ as (x, y). Then

$$f_1(a) = p_1(\hat{f}(a)) = p_1(x, y) = x,$$

and similarly, $f_2(a) = y$. Hence $\hat{f}(a) = (f_1(a), f_2(a)) = \bar{f}(a)$ for all $a \in A$, giving $\hat{f} = \bar{f}$, as required.

In general, in any category, the map $\bar{f}$ of diagram (5.2) is usually written as (f_1, f_2).

Example 5.1.4 In the category of topological spaces, any two objects X and Y have a product. It is the set $X \times Y$ equipped with the product topology and the standard projection maps. The product topology is deliberately designed so that a function

$$\begin{array}{rcl} A & \to & X \times Y \\ t & \mapsto & (x(t), y(t)) \end{array}$$

is continuous if and only if it is continuous in each coordinate (that is to say, both functions

$$t \mapsto x(t), \qquad t \mapsto y(t)$$

are continuous). This holds for any space A, but the idea is perhaps at its most intuitively appealing when $A = \mathbb{R}$ and we think of t as a time parameter.

A closely related statement is that the product topology is the smallest topology on $X \times Y$ for which the projections are continuous. Here 'smallest' means that for any other topology $\mathcal{T}$ on $X \times Y$ such that p_1 and p_2 are continuous, every subset of $X \times Y$ open in the product topology is also open in $\mathcal{T}$. Thus, to define the product topology, we declare just enough sets to be open that the projections are continuous.

Example 5.1.5   Now let X and Y be vector spaces. We can form their direct sum, $X \oplus Y$, whose elements can be written as either (x, y) or $x + y$ (with $x \in X$ and $y \in Y$), according to taste. There are linear projection maps

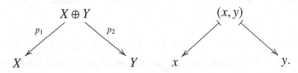

It can be shown that $X \oplus Y$, together with p_1 and p_2, is the product of X and Y in the category of vector spaces (Exercise 5.1.33).

Examples 5.1.6 (Elements of ordered sets)   (a) Let $x, y \in \mathbb{R}$. Their minimum $\min\{x, y\}$ satisfies

$$\min\{x, y\} \leq x, \qquad \min\{x, y\} \leq y$$

and has the further property that whenever $a \in \mathbb{R}$ with

$$a \leq x, \qquad a \leq y,$$

we have $a \leq \min\{x, y\}$. This means exactly that when the poset $(\mathbb{R}, \leq)$ is viewed as a category, the product of $x, y \in \mathbb{R}$ is $\min\{x, y\}$. The definition of product simplifies when interpreted in a poset, since all diagrams commute.

(b) Fix a set S. Let $X, Y \in \mathscr{P}(S)$. Then $X \cap Y$ satisfies

$$X \cap Y \subseteq X, \qquad X \cap Y \subseteq Y$$

and has the further property that whenever $A \in \mathscr{P}(S)$ with

$$A \subseteq X, \qquad A \subseteq Y,$$

we have $A \subseteq X \cap Y$. This means that $X \cap Y$ is the product of X and Y in the poset $(\mathscr{P}(S), \subseteq)$ regarded as a category.

(c) Let $x, y \in \mathbb{N}$. Their greatest common divisor $\gcd(x, y)$ satisfies

$$\gcd(x, y) \mid x, \qquad \gcd(x, y) \mid y$$

(it's a common divisor!) and has the further property that whenever $a \in \mathbb{N}$ with

$$a \mid x, \qquad a \mid y,$$

we have $a \mid \gcd(x, y)$. This means that $\gcd(x, y)$ is the product of x and y in the poset $(\mathbb{N}, \mid)$ regarded as a category.

Generally, let $(A, \leq)$ be a poset and $x, y \in A$. A **lower bound** for x and y is an element $a \in A$ such that $a \leq x$ and $a \leq y$. A **greatest lower bound** or **meet** of x and y is a lower bound z for x and y with the further property that whenever a is a lower bound for x and y, we have $a \leq z$.

When a poset is regarded as a category, meets are exactly products. They do not always exist, but when they do, they are unique. The meet of x and y is usually written as $x \wedge y$ rather than $x \times y$. Thus, in the three examples above,

$$x \wedge y = \min\{x, y\}, \qquad X \wedge Y = X \cap Y, \qquad x \wedge y = \gcd(x, y),$$

the second example being the origin of the notation.

We have been discussing products $X \times Y$ of *two* objects, so-called **binary products**. But there is no reason to stick to two. We can just as well talk about products $X \times Y \times Z$ of three objects, or of infinitely many objects. The definition changes in the most obvious way:

Definition 5.1.7 Let $\mathscr{A}$ be a category, I a set, and $(X_i)_{i \in I}$ a family of objects of $\mathscr{A}$. A **product** of $(X_i)_{i \in I}$ consists of an object P and a family of maps

$$\left(P \xrightarrow{p_i} X_i \right)_{i \in I}$$

with the property that for all objects A and families of maps

$$\left(A \xrightarrow{f_i} X_i \right)_{i \in I} \tag{5.3}$$

there exists a unique map $\bar{f} \colon A \to P$ such that $p_i \circ \bar{f} = f_i$ for all $i \in I$.

Remarks 5.1.2 apply equally to this definition. When the product P exists, we write P as $\prod_{i \in I} X_i$ and the map $\bar{f}$ as $(f_i)_{i \in I}$. We call the maps f_i the **components** of the map $(f_i)_{i \in I}$. Taking I to be a two-element set, we recover the special case of binary products.

Example 5.1.8 In ordered sets, the extension from binary to arbitrary products works in the obvious way: given an ordered set $(A, \leq)$, a **lower bound** for a family $(x_i)_{i \in I}$ of elements is an element $a \in A$ such that $a \leq x_i$ for all i, and a **greatest lower bound** or **meet** of the family is a lower bound greater than any other, written as $\bigwedge_{i \in I} x_i$. These are the products in $(A, \leq)$.

For example, in $\mathbb{R}$ with its usual ordering, the meet of a family $(x_i)_{i \in I}$ is $\inf\{x_i \mid i \in I\}$ (and one exists if and only if the other does).

Example 5.1.9 What happens to the definition of product when the indexing set I is empty? Let $\mathscr{A}$ be a category. In general, an I-indexed family $(X_i)_{i \in I}$ of objects of $\mathscr{A}$ is a function $I \to \mathrm{ob}(\mathscr{A})$. When I is empty, there is exactly one such function. In other words, there is exactly one family $(X_i)_{i \in \emptyset}$, the **empty**

family. Similarly, when I is empty, there is exactly one family (5.3) for any given object A.

A product of the empty family therefore consists of an object P of $\mathscr{A}$ such that for each object A of $\mathscr{A}$, there exists a unique map $\bar{f} \colon A \to P$. (The condition '$p_i \circ \bar{f} = f_i$ for all $i \in I$' holds trivially.) In other words, a product of the empty family is exactly a terminal object.

We have been writing 1 for terminal objects, which was justified by the fact that in categories such as **Set**, **Top**, **Ring** and **Grp**, the terminal object has one element. But we have just seen that the terminal object is the product of no things, which in the context of elementary arithmetic is the number 1. This is a second, related, reason for the notation.

Example 5.1.10   Take an object X of a category $\mathscr{A}$, and a set I. There is a constant family $(X)_{i \in I}$. Its product $\prod_{i \in I} X$, if it exists, is written as X^I and called a **power** of X.

We met powers in **Set** in Section 3.1. When X is a set, X^I is the set of functions from I to X, also written as $\mathbf{Set}(I, X)$.

Equalizers

To define our second type of limit, we need a preliminary piece of terminology: a **fork** in a category consists of objects and maps

$$A \xrightarrow{\;f\;} X \underset{t}{\overset{s}{\rightrightarrows}} Y \tag{5.4}$$

such that $sf = tf$.

Definition 5.1.11   Let $\mathscr{A}$ be a category and let $X \underset{t}{\overset{s}{\rightrightarrows}} Y$ be objects and maps in $\mathscr{A}$. An **equalizer** of s and t is an object E together with a map $E \xrightarrow{\;i\;} X$ such that

$$E \xrightarrow{\;i\;} X \underset{t}{\overset{s}{\rightrightarrows}} Y$$

is a fork, and with the property that for any fork (5.4), there exists a unique map $\bar{f} \colon A \to E$ such that

$$\begin{array}{ccc} & A & \\ \bar{f} \Big\downarrow & & \searrow^{f} \\ E & \xrightarrow[\;i\;]{} & X \end{array} \tag{5.5}$$

commutes.

Remarks 5.1.2 on products apply to equalizers too.

Example 5.1.12 We have already met equalizers in **Set** (Section 3.1). They really are equalizers in the sense of Definition 5.1.11. Indeed, take sets and functions $X \underset{t}{\overset{s}{\rightrightarrows}} Y$, write

$$E = \{x \in X \mid s(x) = t(x)\},$$

and write $i \colon E \to X$ for the inclusion. Then $si = ti$, so we have a fork, and one can check that it is universal among all forks on s and t.

An equalizer describes the set of solutions of a single equation, but by combining equalizers with products, we can also describe the solution-set of any system of simultaneous equations. Take a set Λ and a family

$$\left(X \underset{t_\lambda}{\overset{s_\lambda}{\rightrightarrows}} Y_\lambda \right)_{\lambda \in \Lambda}$$

of pairs of maps in **Set**. Then the solution-set

$$\{x \in X \mid s_\lambda(x) = t_\lambda(x) \text{ for all } \lambda \in \Lambda\}$$

is the equalizer of the functions

$$X \underset{(t_\lambda)_{\lambda \in \Lambda}}{\overset{(s_\lambda)_{\lambda \in \Lambda}}{\rightrightarrows}} \prod_{\lambda \in \Lambda} X_\lambda$$

(using the notation introduced after Definition 5.1.7). To see this, observe that for $x \in X$,

$$(s_\lambda)_{\lambda \in \Lambda}(x) = (t_\lambda)_{\lambda \in \Lambda}(x) \iff (s_\lambda(x))_{\lambda \in \Lambda} = (t_\lambda(x))_{\lambda \in \Lambda}$$
$$\iff s_\lambda(x) = t_\lambda(x) \text{ for all } \lambda \in \Lambda,$$

as required.

Example 5.1.13 Take continuous maps $X \underset{t}{\overset{s}{\rightrightarrows}} Y$ between topological spaces. We can form their equalizer E in the category of sets, with inclusion map $i \colon E \to X$, say. Since E is a subset of the space X, it acquires the subspace topology from X, and i is then continuous. This space E, together with i, is the equalizer of s and t.

Showing this amounts to showing that for any fork (5.4) in **Top**, the induced function $\bar{f}$ is continuous. This follows from the definition of the subspace topology, which is the smallest topology such that the inclusion map is continuous. Compare the remarks on products in Example 5.1.4.

Example 5.1.14 Let $\theta\colon G \to H$ be a homomorphism of groups. As in Example 0.8, the homomorphism θ gives rise to a fork

$$\ker\theta \overset{\iota}{\longrightarrow} G \underset{\varepsilon}{\overset{\theta}{\rightrightarrows}} H$$

where ι is the inclusion and ε is the trivial homomorphism. This is an equalizer in **Grp**. Showing this amounts to showing that the map that we have been calling $\bar{f}$ is a homomorphism, which is left to the reader.

Thus, kernels are a special case of equalizers.

Example 5.1.15 Let $V \underset{t}{\overset{s}{\rightrightarrows}} W$ be linear maps between vector spaces. There is a linear map $t - s\colon V \to W$, and the equalizer of s and t in the category of vector spaces is the space $\ker(t - s)$ together with the inclusion map $\ker(t - s) \hookrightarrow V$.

Pullbacks

We explore one more type of limit before formulating the general definition.

Definition 5.1.16 Let $\mathscr{A}$ be a category, and take objects and maps

$$\begin{array}{c} Y \\ \downarrow{\scriptstyle t} \\ X \xrightarrow{\ s\ } Z \end{array} \qquad (5.6)$$

in $\mathscr{A}$. A **pullback** of this diagram is an object $P \in \mathscr{A}$ together with maps $p_1\colon P \to X$ and $p_2\colon P \to Y$ such that

$$\begin{array}{ccc} P & \xrightarrow{\ p_2\ } & Y \\ {\scriptstyle p_1}\downarrow & & \downarrow{\scriptstyle t} \\ X & \xrightarrow{\ s\ } & Z \end{array} \qquad (5.7)$$

commutes, and with the property that for any commutative square

$$\begin{array}{ccc} A & \xrightarrow{\ f_2\ } & Y \\ {\scriptstyle f_1}\downarrow & & \downarrow{\scriptstyle t} \\ X & \xrightarrow{\ s\ } & Z \end{array} \qquad (5.8)$$

in $\mathscr{A}$, there is a unique map $\bar{f}: A \to P$ such that

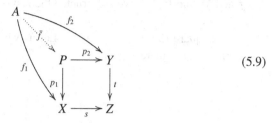

$$(5.9)$$

commutes. (For (5.9) to commute means only that $p_1\bar{f} = f_1$ and $p_2\bar{f} = f_2$, since the commutativity of the square is already given.)

Again, Remarks 5.1.2 apply.

We call (5.7) a **pullback square**. Another name for pullback is **fibred product**. This name is partially explained by the following fact: when Z is a terminal object (and s and t are the only maps they can possibly be), a pullback of the diagram (5.6) is simply a product of X and Y.

Examples 5.1.17 (Pullbacks in Set)   The pullback of a diagram (5.6) in **Set** is

$$P = \{(x, y) \in X \times Y \mid s(x) = t(y)\}$$

with projections p_1 and p_2 given by $p_1(x, y) = x$ and $p_2(x, y) = y$.

Although you might not be familiar with general pullbacks in **Set**, there are at least two instances that you are likely to have met.

(a) A basic construction with sets and functions is the formation of inverse images. They are an instance of pullbacks. Indeed, given a function $f: X \to Y$ and a subset $Y' \subseteq Y$, we obtain a new set, the inverse image

$$f^{-1}Y' = \{x \in X \mid f(x) \in Y'\} \subseteq X,$$

and a new function,

$$f': \quad f^{-1}Y' \quad \to \quad Y'$$
$$x \quad \mapsto \quad f(x).$$

We also have the inclusion functions $j: Y' \hookrightarrow Y$ and $i: f^{-1}Y' \hookrightarrow X$. Putting everything together gives a commutative square

$$\begin{array}{ccc}
f^{-1}Y' & \xrightarrow{f'} & Y' \\
i\downarrow & & \downarrow j \\
X & \xrightarrow{f} & Y.
\end{array}$$

$$(5.10)$$

The data we started with was the lower-right part of this square $(X, Y, Y',$ f and $j)$, and from it we constructed the rest of the square $(f^{-1}Y', f'$ and $i)$.

The square (5.10) is a pullback. Let us verify this in detail. Take any commutative square

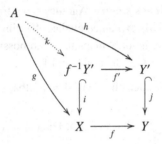

We must show that there is a unique map $k: A \to f^{-1}Y'$ such that

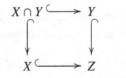

commutes. For uniqueness, let k be a map making the diagram commute. Then for all $a \in A$, we have $i(k(a)) = g(a)$, that is, $k(a) = g(a)$, and this determines k uniquely. For existence, first note that for all $a \in A$ we have $f(g(a)) = j(h(a)) \in Y'$, so $g(a) \in f^{-1}Y'$. Hence we may define $k:$ $A \to f^{-1}Y'$ by $k(a) = g(a)$ for all $a \in A$. Then for all $a \in A$, we have $i(k(a)) = k(a) = g(a)$ and

$$f'(k(a)) = f(k(a)) = f(g(a)) = j(h(a)) = h(a).$$

Hence $i \circ k = g$ and $f' \circ k = h$, as required.

(b) Intersection of subsets provides another example of pullbacks. Indeed, let X and Y be subsets of a set Z. Then

$$
\begin{array}{ccc}
X \cap Y & \hookrightarrow & Y \\
\cap\downarrow & & \cap\downarrow \\
X & \hookrightarrow & Z
\end{array}
$$

is a pullback square, where all the arrows are inclusions of subsets.

In fact, this is a special case of (a), since $X \cap Y$ is the inverse image of $Y \subseteq Z$ under the inclusion map $X \hookrightarrow Z$.

In the situation of Example 5.1.17(a), where we have a map $f: X \to Y$ and a subset Y' of Y, people sometimes say that $f^{-1}Y'$ is obtained by 'pulling Y' back' along f: hence the name.

The definition of limit

We have now looked at three constructions: products, equalizers and pullbacks. They clearly have something in common. Each starts with some objects and (in the case of equalizers and pullbacks) some maps between them. In each, we aim to construct a new object together with some maps from it to the original objects, with a universal property.

Let us analyse this more closely. What is the starting data in each construction? For (binary) products, it is a pair of objects

$$X \qquad Y. \tag{5.11}$$

For equalizers, it is a diagram

$$X \overset{s}{\underset{t}{\rightrightarrows}} Y. \tag{5.12}$$

For pullbacks, it is a diagram

$$
\begin{array}{c}
Y \\
\downarrow{\scriptstyle t} \\
X \xrightarrow{\;s\;} Z.
\end{array}
\tag{5.13}
$$

In Definition 4.1.25, we met the notion of generalized element, and we saw there that the 'figures' in a geometric object can often be described by maps into it. For instance, a curve in a topological space A can be thought of as a map $\mathbb{R} \to A$. Similarly, an object of a category $\mathscr{A}$ amounts to a functor $D: \mathbf{1} \to \mathscr{A}$; think of $\mathbf{1} = \boxed{\bullet}$ as an unlabelled object and D as labelling it with the name of an object of $\mathscr{A}$. And similarly again, a map in a category $\mathscr{A}$ is a functor $\mathbf{2} \to \mathscr{A}$, where $\mathbf{2} = \boxed{\bullet \to \bullet}$. (Here $\mathbf{2}$ is the category with two objects, say 0 and 1, with one map $0 \to 1$, and with no other maps except for identities.) Finally, if we take $\mathbf{I}$ to be one of the categories

$$\mathbf{T} = \boxed{\bullet \qquad \bullet}, \quad \mathbf{E} = \boxed{\bullet \rightrightarrows \bullet} \quad \text{or} \quad \mathbf{P} = \boxed{\begin{array}{c} \bullet \\ \downarrow \\ \bullet \longrightarrow \bullet \end{array}} \tag{5.14}$$

then a functor $\mathbf{I} \to \mathscr{A}$ consists of data (5.11), (5.12) or (5.13) in $\mathscr{A}$, respectively.

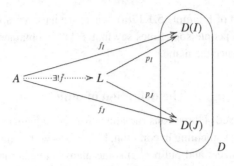

Figure 5.1 The definition of limit.

We have just begun to use the convention that one typeface ($\mathbf{A}$, $\mathbf{B}$, $\mathbf{C}$, ...) denotes small categories, and another ($\mathscr{A}$, $\mathscr{B}$, $\mathscr{C}$, ...) denotes arbitrary categories. Although not strictly necessary, this convention is helpful, since small categories and arbitrary categories often play different roles in the theory.

Definition 5.1.18 Let $\mathscr{A}$ be a category and $\mathbf{I}$ a small category. A functor $\mathbf{I} \to \mathscr{A}$ is called a **diagram** in $\mathscr{A}$ of **shape $\mathbf{I}$**.

So (5.11), (5.12) and (5.13) are diagrams of shape $\mathbf{T}$, $\mathbf{E}$ and $\mathbf{P}$.

We already have the definitions of product of a diagram of shape $\mathbf{T}$, equalizer of a diagram of shape $\mathbf{E}$, and pullback of a diagram of shape $\mathbf{P}$. We now unify them in the definition of limit (Figure 5.1).

Definition 5.1.19 Let $\mathscr{A}$ be a category, $\mathbf{I}$ a small category, and $D\colon \mathbf{I} \to \mathscr{A}$ a diagram in $\mathscr{A}$.

(a) A **cone** on D is an object $A \in \mathscr{A}$ (the **vertex** of the cone) together with a family

$$\left(A \xrightarrow{f_I} D(I)\right)_{I\in\mathbf{I}} \tag{5.15}$$

of maps in $\mathscr{A}$ such that for all maps $I \xrightarrow{u} J$ in $\mathbf{I}$, the triangle

$$
\begin{array}{ccc}
 & & D(I) \\
 & \nearrow^{f_I} & \\
A & & \downarrow Du \\
 & \searrow_{f_J} & \\
 & & D(J)
\end{array}
$$

commutes. (Here and later, we abbreviate $D(u)$ as Du.)

(b) A **limit** of D is a cone $\left(L \xrightarrow{p_I} D(I)\right)_{I\in\mathbf{I}}$ with the property that for any cone (5.15) on D, there exists a unique map $\bar{f}\colon A \to L$ such that $p_I \circ \bar{f} = f_I$ for all $I \in \mathbf{I}$. The maps p_I are called the **projections** of the limit.

Remarks 5.1.20 (a) Loosely, the universal property says that for any $A \in \mathscr{A}$, maps $A \to L$ correspond one-to-one with cones on D with vertex A. (Any map $g: A \to L$ gives rise to a cone $\left(A \xrightarrow{p_I g} D(I)\right)_{I \in \mathbf{I}}$, and the definition of limit is that for each A, this process is bijective.) In Section 6.1, we will use this thought to rephrase the definition of limit in terms of representability. From this it will follow that limits are unique up to canonical isomorphism, when they exist (Corollary 6.1.2). Alternatively, uniqueness can be proved by the usual kind of direct argument, as in Lemma 2.1.8.

(b) If $\left(L \xrightarrow{p_I} D(I)\right)_{I \in \mathbf{I}}$ is a limit of D, we sometimes abuse language slightly by referring to L (rather than the whole cone) as the limit of D. For emphasis, we sometimes call $\left(L \xrightarrow{p_I} D(I)\right)_{I \in \mathbf{I}}$ a **limit cone**. We write $L = \varprojlim_{\mathbf{I}} D$. Remark (a) can then be stated as:

A map into $\varprojlim_{\mathbf{I}} D$ *is a cone on* D.

(c) By assuming from the outset that the shape category $\mathbf{I}$ is small, we are restricting ourselves to what are officially called **small limits**. We will seldom be interested in any other kind.

Examples 5.1.21 (Limit shapes) Let $\mathscr{A}$ be any category. Recall the categories $\mathbf{T}$, $\mathbf{E}$ and $\mathbf{P}$ of (5.14).

(a) A diagram D of shape $\mathbf{T}$ in $\mathscr{A}$ is a pair (X, Y) of objects of $\mathscr{A}$. A cone on D is an object A together with maps $f_1: A \to X$ and $f_2: A \to Y$ (as in Definition 5.1.1), and a limit of D is a product of X and Y.

More generally, let I be a set and write $\mathbf{I}$ for the discrete category on I. A functor $D: \mathbf{I} \to \mathscr{A}$ is an I-indexed family $(X_i)_{i \in I}$ of objects of $\mathscr{A}$, and a limit of D is exactly a product of the family $(X_i)_{i \in I}$.

In particular, a limit of the unique functor $\emptyset \to \mathscr{A}$ is a terminal object of $\mathscr{A}$, where $\emptyset$ denotes the empty category.

(b) A diagram D of shape $\mathbf{E}$ in $\mathscr{A}$ is a parallel pair $X \overset{s}{\underset{t}{\rightrightarrows}} Y$ of maps in $\mathscr{A}$. A cone on D consists of objects and maps

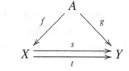

such that $s \circ f = g$ and $t \circ f = g$. But since g is determined by f, it is equivalent to say that a cone on D consists of an object A and a map f:

$A \to X$ such that

$$A \xrightarrow{\ f\ } X \underset{t}{\overset{s}{\rightrightarrows}} Y$$

is a fork. A limit of D is a universal fork on s and t, that is, an equalizer of s and t.

(c) A diagram D of shape $\mathbf{P}$ in $\mathscr{A}$ consists of objects and maps

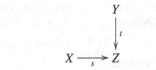

in $\mathscr{A}$. Performing a simplification similar to that in (b), we see that a cone on D is a commutative square (5.8). A limit of D is a pullback.

(d) Let $\mathbf{I} = (\mathbb{N}, \leq)^{\mathrm{op}}$. A diagram $D \colon \mathbf{I} \to \mathscr{A}$ consists of objects and maps

$$\cdots \xrightarrow{s_3} X_2 \xrightarrow{s_2} X_1 \xrightarrow{s_1} X_0.$$

For example, suppose that we have a set X_0 and a chain of subsets

$$\cdots \subseteq X_2 \subseteq X_1 \subseteq X_0.$$

The inclusion maps form a diagram in **Set** of the type above, and its limit is $\bigcap_{i \in \mathbb{N}} X_i$. In this and similar contexts, limits are sometimes referred to as **inverse limits**, although many category theorists regard this usage as old-fashioned.

In general, the limit of a diagram D is the terminal object in the category of cones on D, and is therefore an extremal example of a cone on D. The word 'limit' can be understood as meaning 'on the boundary', rather than indicating a limiting process of the type encountered in analysis. Nevertheless, the two ideas make contact in Example 5.1.21(d).

We have said little so far about which limits exist, except to observe in Remark 5.1.2(a) that they do not exist always. We now show that in many familiar categories, all limits do exist; indeed, we can construct them explicitly.

Example 5.1.22 Let $D \colon \mathbf{I} \to \mathbf{Set}$ and, as a kind of thought experiment, let us ask ourselves what $\lim_{\leftarrow \mathbf{I}} D$ would have to be if it existed. (We do not know yet that it does.) We would have

$$\lim_{\leftarrow \mathbf{I}} D \cong \mathbf{Set}\left(1, \lim_{\leftarrow \mathbf{I}} D\right)$$

$$\cong \{\text{cones on } D \text{ with vertex } 1\}$$

$$\cong \left\{ (x_I)_{I \in \mathbf{I}} \mid x_I \in D(I) \text{ for all } I \in \mathbf{I} \text{ and } (Du)(x_I) = x_J \right.$$

$$\left. \text{for all } I \xrightarrow{u} J \text{ in } \mathbf{I} \right\}, \tag{5.16}$$

where the second isomorphism is by Remark 5.1.20(a) and the third is by definition of cone. In fact, (5.16) really *is* the limit of D in **Set**, with projections $p_J \colon \lim_{\leftarrow \mathbf{I}} D \to D(J)$ given by $p_J((x_I)_{I \in \mathbf{I}}) = x_J$ (Exercise 5.1.37). So in **Set**, all limits exist.

Example 5.1.23 The same formula gives limits in categories of algebras such as **Grp**, **Ring**, **Vect**$_k$, ... Of course, we also have to say what the group/ring/... structure on the set (5.16) is, but this works in the most straightforward way imaginable. For instance, in **Vect**$_k$, if $(x_I)_{I \in \mathbf{I}}, (y_I)_{I \in \mathbf{I}} \in \lim_{\leftarrow \mathbf{I}} D$ then

$$(x_I)_{I \in \mathbf{I}} + (y_I)_{I \in \mathbf{I}} = (x_I + y_I)_{I \in \mathbf{I}}.$$

Example 5.1.24 The same formula also gives limits in **Top**. The topology on the set (5.16) is the smallest for which the projection maps are continuous.

Definition 5.1.25 (a) Let $\mathbf{I}$ be a small category. A category $\mathscr{A}$ **has limits of shape** $\mathbf{I}$ if for every diagram D of shape $\mathbf{I}$ in $\mathscr{A}$, a limit of D exists.

(b) A category **has all limits** (or properly, **has small limits**) if it has limits of shape $\mathbf{I}$ for all small categories $\mathbf{I}$.

Thus, **Set**, **Top**, **Grp**, **Ring**, **Vect**$_k$, ... all have all limits.

Similar terminology can be applied to special classes of limits (for instance, 'has pullbacks'). The class of finite limits is particularly important. By definition, a category is **finite** if it contains only finitely many maps (in which case it also contains only finitely many objects). A **finite limit** is a limit of shape $\mathbf{I}$ for some finite category $\mathbf{I}$. For instance, binary products, terminal objects, equalizers and pullbacks are all finite limits.

The next result tells us that all limits can be built up from limits of just a few familiar, basic types.

Proposition 5.1.26 *Let $\mathscr{A}$ be a category.*

(a) *If $\mathscr{A}$ has all products and equalizers then $\mathscr{A}$ has all limits.*

(b) *If $\mathscr{A}$ has binary products, a terminal object and equalizers then $\mathscr{A}$ has finite limits.*

To understand the idea, consider formula (5.16) for limits in **Set**. There, the limit of a diagram D is described as the subset of the product $\prod_{I \in \mathbf{I}} D(I)$ consisting of those elements for which certain equations hold. We saw in Example 5.1.12 that the set of solutions to any system of simultaneous equations can be described via products and equalizers. Thus, we can describe any limit in **Set** in terms of products and equalizers. And in fact, this same description is valid in any category.

We now examine this idea more closely, in preparation for the proof (Exercise 5.1.38). First-time readers may wish to skip the next two paragraphs, resuming at Example 5.1.27.

Equation (5.16) states that in **Set**, the limit of a diagram $D \colon \mathbf{I} \to \mathbf{Set}$ consists of the elements $(x_I)_{I \in \mathbf{I}} \in \prod_{I \in \mathbf{I}} D(I)$ such that

$$(Du)(x_J) = x_K$$

in $D(K)$ for each map $J \xrightarrow{u} K$ in $\mathbf{I}$. For each such map u, define maps

$$\prod_{I \in \mathbf{I}} D(I) \overset{s_u}{\underset{t_u}{\rightrightarrows}} D(K)$$

by

$$s_u((x_I)_{I \in \mathbf{I}}) = (Du)(x_J), \qquad t_u((x_I)_{I \in \mathbf{I}}) = x_K.$$

Then $\lim_{\leftarrow \mathbf{I}} D$ is the set of families $x = (x_I)_{I \in \mathbf{I}}$ satisfying the equation $s_u(x) = t_u(x)$ for each map u in $\mathbf{I}$. It follows from Example 5.1.12 that $\lim_{\leftarrow \mathbf{I}} D$ is the equalizer of

$$\prod_{I \in \mathbf{I}} D(I) \overset{s}{\underset{t}{\rightrightarrows}} \prod_{J \xrightarrow{u} K \text{ in } \mathbf{I}} D(K)$$

where s and t are the maps with components s_u and t_u, respectively.

We have now described any limit in **Set** in terms of products and equalizers. Although our argument took place entirely in **Set**, it suggests how we might proceed in an arbitrary category. With this in mind, the proof of Proposition 5.1.26 is routine, and is left as Exercise 5.1.38.

Example 5.1.27   Let **CptHff** denote the category of compact Hausdorff spaces and continuous maps. It is a classic exercise in topology to show that given continuous maps s and t from a topological space X to a Hausdorff space Y, the subset $\{x \in X \mid s(x) = t(x)\}$ of X is closed. From this it follows that **CptHff** has equalizers. Also, Tychonoff's theorem states that any product (in **Top**) of compact spaces is compact, and it is easy to show that any product (in

Top) of Hausdorff spaces is Hausdorff. From this it follows that **CptHff** has all products. Hence by Proposition 5.1.26(a), **CptHff** has all limits.

Example 5.1.28 Recall from Example 5.1.15 that kernels provide equalizers in **Vect**$_k$. By Proposition 5.1.26(b), finite limits in **Vect**$_k$ can always be expressed in terms of $\oplus$ (binary direct sum), $\{0\}$, and kernels. The same is true in **Ab**.

Monics

For functions between sets, injectivity is an important concept. For maps in an arbitrary category, injectivity does not make sense, but there is a concept that plays a similar role.

Definition 5.1.29 Let $\mathscr{A}$ be a category. A map $X \xrightarrow{f} Y$ in $\mathscr{A}$ is **monic** (or a **monomorphism**) if for all objects A and maps $A \underset{x'}{\overset{x}{\rightrightarrows}} X$,

$$f \circ x = f \circ x' \implies x = x'.$$

This can be rephrased suggestively in terms of generalized elements: f is monic if for all generalized elements x and x' of X (of the same shape), $fx = fx' \implies x = x'$. Being monic is, therefore, the generalized-element analogue of injectivity.

Example 5.1.30 In **Set**, a map is monic if and only if it is injective. Indeed, if f is injective then certainly f is monic, and for the converse, take $A = 1$.

Example 5.1.31 In categories of algebras such as **Grp**, **Vect**$_k$, **Ring**, etc., it is also true that the monic maps are exactly the injections. Again, it is easy to show that injections are monic. For the converse, take $A = F(1)$ where F is the free functor (Examples 2.1.3).

Why is the definition of monic in a chapter on limits? Because of this:

Lemma 5.1.32 *A map $X \xrightarrow{f} Y$ is monic if and only if the square*

$$\begin{array}{ccc} X & \xrightarrow{1} & X \\ {\scriptstyle 1}\downarrow & & \downarrow{\scriptstyle f} \\ X & \xrightarrow[f]{} & Y \end{array}$$

is a pullback.

Proof Exercise 5.1.41. □

The significance of this lemma is that whenever we prove a result about limits, a result about monics will follow. For example, we will soon show that the forgetful functors from **Grp**, **Vect**$_k$, etc., to **Set** preserve limits (in a sense to be defined), from which it will follow immediately that they also preserve monics. This in turn gives an alternative proof that monics in these categories are injective.

Exercises

5.1.33   Verify that in the category of vector spaces, the product of two vector spaces is their direct sum (Example 5.1.5).

5.1.34   Take objects and maps $E \xrightarrow{\;i\;} X \underset{g}{\overset{f}{\rightrightarrows}} Y$ in some category. If this is an equalizer, is the square

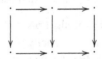

necessarily a pullback? What about the converse? Give proofs or counterexamples.

5.1.35   Take a commutative diagram

$$
\begin{array}{ccccc}
\bullet & \longrightarrow & \bullet & \longrightarrow & \bullet \\
\downarrow & & \downarrow & & \downarrow \\
\bullet & \longrightarrow & \bullet & \longrightarrow & \bullet
\end{array}
$$

in some category. Suppose that the right-hand square is a pullback. Show that the left-hand square is a pullback if and only if the outer rectangle is a pullback.

5.1.36   Let $D \colon \mathbf{I} \to \mathscr{A}$ be a diagram and $\left(L \xrightarrow{p_I} D(I) \right)_{I \in \mathbf{I}}$ a limit cone on D.

(a) Prove that whenever $A \underset{h'}{\overset{h}{\rightrightarrows}} L$ are maps such that $p_I \circ h = p_I \circ h'$ for all $I \in \mathbf{I}$, then $h = h'$.

(b) What does the result of (a) mean when $\mathbf{I}$ is the two-object discrete category, $\mathscr{A} = \mathbf{Set}$, and $A = 1$? Answer without using any category-theoretic terminology.

5.1.37   Show that the set (5.16) in Example 5.1.22 really is a limit of D.

5.1.38 In this exercise, you will prove Proposition 5.1.26, following the plan described after the statement of that proposition.

(a) Let $\mathscr{A}$ be a category with all products and equalizers. Let $D \colon \mathbf{I} \to \mathscr{A}$ be a diagram in $\mathscr{A}$. Define maps

$$\prod_{I \in \mathbf{I}} D(I) \underset{t}{\overset{s}{\rightrightarrows}} \prod_{J \overset{u}{\to} K \text{ in } \mathbf{I}} D(K)$$

as follows: given $J \overset{u}{\to} K$ in $\mathbf{I}$, the u-component of s is the composite

$$\prod_{I \in \mathbf{I}} D(I) \overset{\mathrm{pr}_J}{\to} D(J) \overset{Du}{\to} D(K)$$

(where pr denotes a product projection), and the u-component of t is pr_K. Let $L \overset{p}{\to} \prod_{I \in \mathbf{I}} D(I)$ be the equalizer of s and t, and write p_I for the I-component of p. Show that $\left(L \overset{p_I}{\to} D(I) \right)_{I \in \mathbf{I}}$ is a limit cone on D, thus proving Proposition 5.1.26(a).

(b) Adapt the argument to prove Proposition 5.1.26(b).

5.1.39 Prove that a category with pullbacks and a terminal object has all finite limits.

5.1.40 Let $\mathscr{A}$ be a category and $A \in \mathscr{A}$. A **subobject** of A is an isomorphism class of monics into A. More precisely, let **Monic**(A) be the full subcategory of $\mathscr{A}/A$ whose objects are the monics; then a subobject of A is an isomorphism class of objects of **Monic**(A).

(a) Let $X \overset{m}{\to} A$ and $X' \overset{m'}{\to} A$ be monics in **Set**. Show that m and m' are isomorphic in **Monic**(A) if and only if they have the same image. Deduce that the subobjects of A are in canonical one-to-one correspondence with the subsets of A.

(b) Part (a) says that in **Set**, subobjects are subsets. What are subobjects in **Grp**, **Ring** and **Vect**$_k$?

(c) What are subobjects in **Top**? (Careful!)

5.1.41 Prove Lemma 5.1.32.

5.1.42 Let

$$\begin{array}{ccc} X' & \overset{f'}{\longrightarrow} & X \\ \scriptstyle{m'}\downarrow & & \downarrow\scriptstyle{m} \\ A' & \underset{f}{\longrightarrow} & A \end{array}$$

be a pullback square in some category. Show that if m is monic then so is m'. (We already know this in the category of sets, by Example 5.1.17(a).)

5.2 Colimits: definition and examples

We have seen that examples of limits occur throughout mathematics. It therefore makes sense to examine the dual concept, colimit, and ask whether it is similarly ubiquitous.

By dualizing, we can write down the definition of colimit immediately. We then specialize to sums, coequalizers and pushouts, the duals of products, equalizers and pullbacks.

There are two common conventions for naming dual concepts: sometimes we add or subtract the prefix 'co' (as in limit/colimit), and sometimes we use 'left' and 'right' (as for adjoints). There are also some irregular names, such as terminal/initial object and pullback/pushout.

Definition 5.2.1 Let $\mathscr{A}$ be a category and $\mathbf{I}$ a small category. Let $D\colon \mathbf{I} \to \mathscr{A}$ be a diagram in $\mathscr{A}$, and write D^{op} for the corresponding functor $\mathbf{I}^{\mathrm{op}} \to \mathscr{A}^{\mathrm{op}}$. A **cocone** on D is a cone on D^{op}, and a **colimit** of D is a limit of D^{op}.

Explicitly, a cocone on D is an object $A \in \mathscr{A}$ (the **vertex** of the cocone) together with a family

$$\left(D(I) \xrightarrow{\ f_I\ } A \right)_{I \in \mathbf{I}} \tag{5.17}$$

of maps in $\mathscr{A}$ such that for all maps $I \xrightarrow{\ u\ } J$ in $\mathbf{I}$, the diagram

$$
\begin{array}{c}
D(I) \\
{\scriptstyle Du}\downarrow \quad\searrow{\scriptstyle f_I} \\
\quad\quad\quad A \\
D(J) \quad \nearrow{\scriptstyle f_J}
\end{array}
$$

commutes. A colimit of D is a cocone

$$\left(D(I) \xrightarrow{\ p_I\ } C \right)_{I \in \mathbf{I}}$$

with the property that for any cocone (5.17) on D, there is a unique map $\bar{f}\colon C \to A$ such that $\bar{f} \circ p_I = f_I$ for all $I \in \mathbf{I}$. The associated picture is the mirror image of Figure 5.1.

Of course, Remarks 5.1.20 apply equally here. We write (the vertex of) the colimit as $\varinjlim_{\mathbf{I}} D$, and call the maps p_I **coprojections**.

Sums

Definition 5.2.2 A **sum** or **coproduct** is a colimit over a discrete category. (That is, it is a colimit of shape **I** for some discrete category **I**.)

Let $(X_i)_{i \in I}$ be a family of objects of a category. Their sum (if it exists) is written as $\sum_{i \in I} X_i$ or $\coprod_{i \in I} X_i$. When I is a finite set $\{1, \ldots, n\}$, we write $\sum_{i \in I} X_i$ as $X_1 + \cdots + X_n$, or as 0 if $n = 0$.

Example 5.2.3 By the dual of Example 5.1.9, a sum of the empty family is exactly an initial object.

Example 5.2.4 Sums in **Set** were described in Section 3.1. Let us look in detail at the universal property, in the case of binary sums. Take two sets, X_1 and X_2. Form their sum, $X_1 + X_2$, and consider the inclusions

$$X_1 \xrightarrow{\ p_1\ } X_1 + X_2 \xleftarrow{\ p_2\ } X_2.$$

This is a colimit cocone. To prove this, we have to prove the following universal property: for any diagram

$$X_1 \xrightarrow{\ f_1\ } A \xleftarrow{\ f_2\ } X_2$$

of sets and functions, there is a unique function $\bar{f} \colon X_1 + X_2 \to A$ making

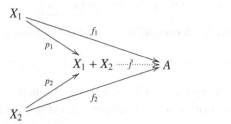

commute. Now, we noted in Section 3.1 that p_1 and p_2 are injections whose images partition $X_1 + X_2$. This means that every element x of $X_1 + X_2$ is *either* equal to $p_1(x_1)$ for some $x_1 \in X_1$ (and this x_1 is then unique), *or* equal to $p_2(x_2)$ for some $x_2 \in X_2$ (and this x_2 is then unique), but not both. So we may define $\bar{f}(x)$ to be equal to $f_1(x_1)$ in the first case and $f_2(x_2)$ in the second. This defines a function $\bar{f}$ making the diagram commute, and it is clearly the unique function that does so.

Example 5.2.5 Let X_1 and X_2 be vector spaces. There are linear maps

$$X_1 \xrightarrow{\ i_1\ } X_1 \oplus X_2 \xleftarrow{\ i_2\ } X_2 \tag{5.18}$$

defined by $i_1(x_1) = (x_1, 0)$ and $i_2(x_2) = (0, x_2)$, and it can be checked that (5.18) is a colimit cocone in $\mathbf{Vect}_k$. Hence binary direct sums are sums in the categorical sense. This is remarkable, since we saw in Example 5.1.5 that $X_1 \oplus X_2$ is also the *product* of X_1 and X_2! Contrast this with the category of sets (or almost any other category), where sums and products are very different.

Example 5.2.6 Let $(A, \le)$ be an ordered set. **Upper bounds** and **least upper bounds** (or **joins**) in A are defined by dualizing the definitions in Example 5.1.6, and, dually, they are sums in the corresponding category. The join of a family $(x_i)_{i \in I}$ is written as $\bigvee_{i \in I} x_i$. In the binary case (where I has two elements), the join of x_1 and x_2 is written as $x_1 \vee x_2$. A join of the empty family (where $I = \emptyset$) is an initial object of the category A, as in Example 5.2.3. Equivalently, it is a **least element** of A: an element $0 \in A$ such that $0 \le a$ for all $a \in A$.

For instance, in $(\mathbb{R}, \le)$, join is supremum and there is no least element. In a power set $(\mathscr{P}(S), \subseteq)$, join is union and the least element is $\emptyset$. In $(\mathbb{N}, |)$, join is lowest common multiple and the least element is 1 (since 1 divides everything). So in this order on the natural numbers, 1 is least; but also, everything divides 0, so 0 is greatest!

Coequalizers

We continue to write $\mathbf{E}$ for the category $\boxed{\bullet \rightrightarrows \bullet}$.

Definition 5.2.7 A **coequalizer** is a colimit of shape $\mathbf{E}$.

In other words, given a diagram $X \overset{s}{\underset{t}{\rightrightarrows}} Y$, a coequalizer of s and t is a map $Y \overset{p}{\longrightarrow} C$ satisfying $p \circ s = p \circ t$ and universal with this property.

We will see that coequalizers are something like quotients. But first, we need some background material on equivalence relations.

Remarks 5.2.8 A binary relation R on a set A can be viewed as a subset $R \subseteq A \times A$. Think of $(a, a') \in R$ as meaning 'a and a' are related'. We can speak of one relation S on A 'containing' another such relation, R. This means that $R \subseteq S$: whenever a and a' are R-related, they are also S-related.

We will need to use the fact that for any binary relation R on a set A, there is a smallest equivalence relation $\sim$ containing R. This is called the equivalence relation **generated** by R. 'Smallest' means that any equivalence relation containing R also contains $\sim$.

We can construct $\sim$ as the intersection of all equivalence relations on A containing R, since the intersection of any family of equivalence relations is again an equivalence relation. There is also an explicit construction. The rough idea is as follows: writing $x \to y$ to mean $(x, y) \in R$, we should have $a \sim a'$ if and only if there is a zigzag such as

$$a \to b \leftarrow c \leftarrow d \to e \leftarrow a'$$

between a and a'. To make this precise, we first define a relation S on A by

$$S = \{(a, a') \in A \times A \mid (a, a') \in R \text{ or } (a', a) \in R\}$$

(which enlarges R to a symmetric relation), then define $\sim$ by declaring that $a \sim a'$ if and only if there exist $n \geq 0$ and $a_0, \ldots, a_n \in A$ such that

$$a = a_0, \ (a_0, a_1) \in S, \ (a_1, a_2) \in S, \ \ldots, \ (a_{n-1}, a_n) \in S, \ a_n = a'$$

(which forces reflexivity and transitivity, while preserving the symmetry).

Next, recall some facts about equivalence relations from Section 3.1. Given any equivalence relation $\sim$ on a set A, we can construct the set $A/\sim$ of equivalence classes and the quotient map $p\colon A \to A/\sim$. This quotient map p is surjective and has the property that $p(a) = p(a') \iff a \sim a'$, for $a, a' \in A$. We saw that for any set B, the maps $A/\sim \to B$ correspond one-to-one (via composition with p) with the maps $f\colon A \to B$ such that

$$\forall a, a' \in A, \qquad a \sim a' \implies f(a) = f(a'). \tag{5.19}$$

Finally, let us consider this universal property in the case where $\sim$ is the equivalence relation generated by some relation R. Condition (5.19) is then equivalent to:

$$\forall a, a' \in A, \qquad (a, a') \in R \implies f(a) = f(a'). \tag{5.20}$$

(Proof: define an equivalence relation $\approx$ on A by $a \approx a' \iff f(a) = f(a')$. Condition (5.19) says that $\sim \subseteq \approx$, and condition (5.20) that $R \subseteq \approx$. But $\sim$ is the smallest equivalence relation containing R, so these statements are equivalent.) In conclusion, for any set B, the maps $A/\sim \to B$ correspond one-to-one with the maps $f\colon A \to B$ satisfying (5.20).

Example 5.2.9 Take sets and functions $X \underset{t}{\overset{s}{\rightrightarrows}} Y$. To find the coequalizer of s and t, we must construct in some canonical way a set C and a function $p\colon Y \to C$ such that $p(s(x)) = p(t(x))$ for all $x \in X$. So, let $\sim$ be the equivalence relation on Y generated by $s(x) \sim t(x)$ for all $x \in X$. (In other words, $\sim$ is generated by the relation

$$R = \{(s(x), t(x)) \mid x \in X\}$$

on Y.) Take the quotient map $p\colon Y \to Y/\sim$. By the correspondence described in Remarks 5.2.8, this is indeed the coequalizer of s and t.

Example 5.2.10   For each pair of homomorphisms $A \overset{s}{\underset{t}{\rightrightarrows}} B$ in **Ab**, there is
a homomorphism $t - s \colon A \to B$, which gives rise to a subgroup $\mathrm{im}(t - s)$ of B.
The coequalizer of s and t is the canonical homomorphism $B \to B/\mathrm{im}(t - s)$.
(Compare Example 5.1.15.)

Pushouts

Definition 5.2.11   A **pushout** is a colimit of shape

$$\mathbf{P}^{\mathrm{op}} = \boxed{\begin{array}{ccc} \bullet & \longrightarrow & \bullet \\ \downarrow & & \\ \bullet & & \end{array}} .$$

In other words, the pushout of a diagram

$$\begin{array}{ccc} X & \overset{s}{\longrightarrow} & Y \\ {\scriptstyle t}\downarrow & & \\ Z & & \end{array} \qquad\qquad (5.21)$$

is (if it exists) a commutative square

$$\begin{array}{ccc} X & \overset{s}{\longrightarrow} & Y \\ {\scriptstyle t}\downarrow & & \downarrow \\ Z & \longrightarrow & \cdot \end{array}$$

that is universal as such. In other words still, a pushout in a category $\mathscr{A}$ is a
pullback in $\mathscr{A}^{\mathrm{op}}$.

Example 5.2.12   Take a diagram (5.21) in **Set**. Its pushout P is $(Y + Z)/\sim$,
where $\sim$ is the equivalence relation on $Y + Z$ generated by $s(x) \sim t(x)$ for all
$x \in X$. The coprojection $Y \to P$ sends $y \in Y$ to its equivalence class in P, and
similarly for the coprojection $Z \to P$.

For example, let Y and Z be subsets of some set A. Then

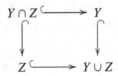

is a pushout square in **Set**. (It is also a pullback square! This coincidence is a
special property of the category of sets.) You can check this by verifying the

universal property or by using the formula just stated. In this case, the formula takes the two sets Y and Z, places them side by side (giving $Y + Z$), then glues the subset $Y \cap Z$ of Y to the subset $Y \cap Z$ of Z (giving $(Y + Z)/\sim\ =\ Y \cup Z$).

Example 5.2.13 If $\mathscr{A}$ is a category with an initial object 0, and if $Y, Z \in \mathscr{A}$, then a pushout of the unique diagram

is exactly a sum of Y and Z.

Example 5.2.14 The van Kampen theorem (Example 0.9) says that given a pushout square in **Top** satisfying certain further hypotheses, the square in **Grp** obtained by taking fundamental groups throughout is also a pushout.

Here is one more shape of colimit, dual to that in Example 5.1.21(d).

Example 5.2.15 A diagram $D \colon (\mathbb{N}, \leq) \to \mathscr{A}$ consists of objects and maps

$$X_0 \xrightarrow{s_1} X_1 \xrightarrow{s_2} X_2 \xrightarrow{s_3} \cdots$$

in $\mathscr{A}$. Colimits of such diagrams are traditionally called **direct limits**. Although the old terms 'inverse limit' (Example 5.1.21(d)) and 'direct limit' are made redundant by the general categorical terms 'limit' and 'colimit' respectively, it is worth being aware of them.

With all these examples in mind, we now write down a general formula for colimits in **Set**.

Example 5.2.16 The colimit of a diagram $D \colon \mathbf{I} \to \mathbf{Set}$ is given by

$$\varinjlim_{\to \mathbf{I}} D = \left(\sum_{I \in \mathbf{I}} D(I) \right) \bigg/ \sim$$

where $\sim$ is the equivalence relation on $\sum D(I)$ generated by

$$x \sim (Du)(x)$$

for all $I \xrightarrow{u} J$ in $\mathbf{I}$ and $x \in D(I)$. To see this, note that for any set A, the maps

$$\left(\sum D(I) \right) \bigg/ \sim\ \to A$$

correspond bijectively with the maps $f \colon \sum D(I) \to A$ such that

$$f(x) = f((Du)(x))$$

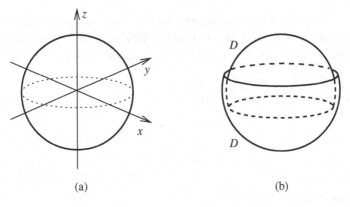

Figure 5.2 Sphere as (a) a limit, and (b) a colimit.

for all u and x (by Remarks 5.2.8). These in turn correspond to families of maps $\left(D(I) \xrightarrow{f_I} A\right)_{I \in \mathbf{I}}$ such that $f_I(x) = f_J((Du)(x))$ for all u and x; but these are exactly the cocones on D with vertex A.

There is a kind of duality between the formulas for limits in **Set** (Example 5.1.22) and colimits in **Set**. Whereas the limit is constructed as a *subset* of a *product*, the colimit is a *quotient* of a *sum*.

Figure 5.2 is intended to convey the difference in flavour between limits and colimits, in a particular topological context. In elementary texts, surfaces are almost always seen as subsets of Euclidean space $\mathbb{R}^3$, with the sphere S^2 typically defined as

$$\{(x, y, z) \in \mathbb{R}^3 \mid x^2 + y^2 + z^2 = 1\}.$$

This is a *subspace* of the *product* space $\mathbb{R}^3 = \mathbb{R} \times \mathbb{R} \times \mathbb{R}$, which suggests that it is a limit. Indeed, the sphere is the equalizer

$$S^2 \lhook\joinrel\longrightarrow \mathbb{R}^3 \overset{s}{\underset{t}{\rightrightarrows}} \mathbb{R}$$

where the maps $s, t \colon \mathbb{R}^3 \to \mathbb{R}$ are given by

$$s(x, y, z) = x^2 + y^2 + z^2, \qquad t(x, y, z) = 1.$$

(An *equa*tion is captured by an *equa*lizer.)

In more advanced mathematics, however, this point of view is used less often. A surface can instead be thought of as the gluing-together of lots of little patches, each isomorphic to the open unit disk D. For example, we could in principle construct an entire bicycle inner tube by gluing together a large number of puncture-repair patches. Figure 5.2(b) shows the simpler example of a sphere made up of two disks glued together. This realizes the sphere as a *quotient* (gluing) of the *sum* (disjoint union) of the two copies of D, suggesting that we have constructed the sphere as a colimit. Indeed, the sphere is the coequalizer

$$S^1 \times (0, 1) \overset{\subset}{\underset{\subset}{\rightrightarrows}} D + D \longrightarrow S^2$$

where S^1 is the circle, the cylinder $S^1 \times (0, 1)$ is the intersection of the two copies of D (the central belt of Figure 5.2(b)), and the two maps into $D + D$ are the inclusions of the cylinder into the first and second copies of D.

One disadvantage of the limit point of view is that it makes an arbitrary choice of coordinate system. It is generally best to think of spaces as free-standing objects, existing independently of any particular embedding into Euclidean space.

One disadvantage of the colimit point of view is that it makes an arbitrary choice of decomposition. For example, we could decompose the sphere into three patches rather than two, or use a different two patches from those shown.

The colimit point of view has the upper hand in modern geometry. (If you are familiar with the definition of manifold, you will recognize that an atlas is essentially a way of viewing a manifold as a colimit of Euclidean balls.) One reason for this is that we are often concerned with maps *out* of spaces X, such as maps $X \to \mathbb{R}$. Maps *out* of a colimit are easy; it is in the very definition of colimit that we know what the maps out of it are.

Epics

Definition 5.2.17 Let $\mathscr{A}$ be a category. A map $X \overset{f}{\longrightarrow} Y$ in $\mathscr{A}$ is **epic** (or an **epimorphism**) if for all objects Z and maps $Y \overset{g}{\underset{g'}{\rightrightarrows}} Z$,

$$g \circ f = g' \circ f \implies g = g'.$$

This is the formal dual of the definition of monic. (In other words, an epic in $\mathscr{A}$ is a monic in $\mathscr{A}^{\mathrm{op}}$.) It is in some sense the categorical version of surjectivity. But whereas the definition of monic closely resembles the definition of injective, the definition of epic does not look much like the definition of surjective. The following examples confirm that in categories where surjectivity makes sense, it is only sometimes equivalent to being epic.

Example 5.2.18 In **Set**, a map is epic if and only if it is surjective. If f is surjective then certainly f is epic. To see the converse, take Z to be a two-element set {`true`, `false`}, take g to be the characteristic function of the image of f (as defined in Section 3.1), and take g' to be the function with constant value `true`.

Any isomorphism in any category is both monic and epic. In **Set**, the converse also holds, since any injective surjective function is invertible (Example 1.1.5).

Example 5.2.19   In categories of algebras, any surjective map is certainly epic. In some such categories, including **Ab**, **Vect**$_k$ and **Grp**, the converse also holds. (The proof is straightforward for **Ab** and **Vect**$_k$, but much harder for **Grp**.) However, there are other categories of algebras where it fails. For instance, in **Ring**, the inclusion $\mathbb{Z} \hookrightarrow \mathbb{Q}$ is epic but not surjective (Exercise 5.2.23). This is also an example of a map that is monic and epic but not an isomorphism.

Example 5.2.20   In the category of Hausdorff topological spaces and continuous maps, any map with dense image is epic.

Of course, there is a dual of Lemma 5.1.32, saying that a map is epic if and only if a certain square is a pushout.

Exercises

5.2.21   Let $X \overset{s}{\underset{t}{\rightrightarrows}} Y$ be maps in some category. Prove that $s = t$ if and only if the equalizer of s and t exists and is an isomorphism, if and only if the coequalizer of s and t exists and is an isomorphism.

5.2.22   (a) Let X be a set and $f \colon X \to X$ a map. Describe the coequalizer of
$$X \overset{f}{\underset{1}{\rightrightarrows}} X$$
in **Set** as explicitly as possible.

(b) Do the same in **Top** rather than **Set**. When X is the circle S^1, find an f such that the coequalizer is an uncountable space with the indiscrete topology.

5.2.23   (a) Prove that in the category of monoids, the inclusion $(\mathbb{N}, +, 0) \hookrightarrow (\mathbb{Z}, +, 0)$ is epic, even though it is not surjective.

(b) Prove that in the category of rings, the inclusion $\mathbb{Z} \hookrightarrow \mathbb{Q}$ is epic, even though it is not surjective.

5.2.24   (Compare Exercise 5.1.40.) Let $\mathscr{A}$ be a category and $A \in \mathscr{A}$. Define a **quotient object** of A to be an isomorphism class of epics out of A. That is, let **Epic**(A) be the full subcategory of $A/\mathscr{A}$ whose objects are the epics; then a quotient object of A is an isomorphism class of objects of **Epic**(A).

(a) Let $A \xrightarrow{e} X$ and $A \xrightarrow{e'} X'$ be epics in **Set**. Show that e and e' are isomorphic in **Epic**(A) if and only if they induce the same equivalence relation on A. Deduce that the quotient objects of A are in canonical one-to-one correspondence with the equivalence relations on A.

(b) Assuming the (nontrivial) fact that the epics in **Grp** are the surjections, show that the quotient objects of a group correspond one-to-one with its normal subgroups.

(The name 'quotient object' is not standard, and indeed there is no standard name for it. Arguably, 'quotient object' would be more suitable for an isomorphism class of *regular* epics, as defined in the following exercises.)

5.2.25 A map $m: A \to B$ is **regular monic** if there exist an object C and maps $B \rightrightarrows C$ of which m is an equalizer. A map $m: A \to B$ is **split monic** if there exists a map $e: B \to A$ such that $em = 1_A$.

(a) Show that split monic $\implies$ regular monic $\implies$ monic.

(b) In **Ab**, show that all monics are regular but not all monics are split. (Hint for the first part: equalizers in **Ab** are calculated as in Example 5.1.15.)

(c) In **Top**, describe the regular monics, and find a monic that is not regular.

5.2.26 Dualizing the definitions in Exercise 5.2.25 gives definitions of **regular** and **split epic**.

(a) We saw in Example 5.2.19 that a map may be monic and epic but not an isomorphism. Prove that in any category, a map is an isomorphism if and only if it is both monic and *regular* epic.

(b) Using the assumption that our category of sets satisfies the axiom of choice (Section 3.1), show that

$$\text{epic} \iff \text{regular epic} \iff \text{split epic}$$

in **Set**.

(c) Let us say that a category $\mathscr{A}$ satisfies the **axiom of choice** if all epics in $\mathscr{A}$ are split. Prove that neither **Top** nor **Grp** satisfies the axiom of choice.

5.2.27 The result of Exercise 5.1.42 can be phrased as 'the class of monics is stable under pullback'. It is also a fact that the composite of two monics is always monic; we say that the class of monics is 'closed under composition'.

Consider the following six classes of map:

monics, regular monics, split monics, epics, regular epics, split epics.

Determine whether each class is stable under pullback or closed under composition.

5.3 Interactions between functors and limits

We saw in Example 5.1.23 that limits in categories such as **Grp**, **Ring** and **Vect**$_k$ can be computed by first taking the limit in the category of sets, then equipping the result with a suitable algebraic structure. On the other hand, colimits in these categories are unlike colimits in **Set**. For example, the underlying set of the initial object of **Grp** (which has one element) is not the initial object of **Set** (which has no elements), and the underlying set of the direct sum $X \oplus Y$ of two vector spaces is not the sum of the underlying sets of X and Y. So, these forgetful functors interact well with limits and badly with colimits.

In this section, we develop terminology that will enable us to express these thoughts precisely.

Definition 5.3.1   (a) Let **I** be a small category. A functor $F \colon \mathscr{A} \to \mathscr{B}$ **preserves limits of shape I** if for all diagrams $D \colon \mathbf{I} \to \mathscr{A}$ and all cones $\left(A \xrightarrow{p_I} D(I) \right)_{I \in \mathbf{I}}$ on D,

$$\left(A \xrightarrow{p_I} D(I) \right)_{I \in \mathbf{I}} \text{ is a limit cone on } D \text{ in } \mathscr{A}$$

$$\implies \left(F(A) \xrightarrow{Fp_I} FD(I) \right)_{I \in \mathbf{I}} \text{ is a limit cone on } F \circ D \text{ in } \mathscr{B}.$$

(b) A functor $F \colon \mathscr{A} \to \mathscr{B}$ **preserves limits** if it preserves limits of shape **I** for all small categories **I**.

(c) **Reflection** of limits is defined as in (a), but with $\impliedby$ in place of $\implies$.

Of course, the same terminology applies to colimits.

Here is a different way to state the definition of preservation. A functor $F \colon \mathscr{A} \to \mathscr{B}$ preserves limits if and only if it has the following property: whenever $D \colon \mathbf{I} \to \mathscr{A}$ is a diagram that has a limit, the composite $F \circ D \colon \mathbf{I} \to \mathscr{B}$ also has a limit, and the canonical map

$$F\left(\lim_{\leftarrow \mathbf{I}} D \right) \to \lim_{\leftarrow \mathbf{I}} (F \circ D)$$

is an isomorphism. Here the 'canonical map' has I-component

$$F\left(\lim_{\leftarrow \mathbf{I}} D \right) \xrightarrow{F(p_I)} F(D(I)),$$

where p_I is the Ith projection of the limit cone on D.

In particular, if F preserves limits then

$$F\left(\lim_{\leftarrow \mathbf{I}} D \right) \cong \lim_{\leftarrow \mathbf{I}} (F \circ D) \tag{5.22}$$

whenever D is a diagram with a limit. Preservation of limits says more than

(5.22) does: the left- and right-hand sides are required to be not just isomorphic, but isomorphic *in a particular way*. Nevertheless, we will sometimes omit this check, acting as if preservation means only that (5.22) holds.

Example 5.3.2 The forgetful functor U: **Top** $\to$ **Set** preserves both limits and colimits. (As we will see, this follows from the fact that U has adjoints on both sides.) It does not reflect all limits or all colimits. For instance, choose any non-discrete spaces X and Y, and let Z be the set $U(X) \times U(Y)$ equipped with the discrete topology. (All that matters here is that the topology on Z is strictly larger than the product topology.) Then we have a cone

$$X \leftarrow Z \to Y \tag{5.23}$$

in **Top** whose image in **Set** is the product cone

$$U(X) \leftarrow U(X) \times U(Y) \to U(Y).$$

But (5.23) is not a product cone in **Top**, since the discrete topology on $U(X) \times U(Y)$ is not the product topology.

Example 5.3.3 In the first paragraph of this section, we observed that the forgetful functor **Grp** $\to$ **Set** does not preserve initial objects and that the forgetful functor **Vect**$_k$ $\to$ **Set** does not preserve binary sums. Forgetful functors out of categories of algebras very seldom preserve all colimits.

Example 5.3.4 We also saw that (in the examples mentioned) forgetful functors on categories of algebras do preserve limits. In fact, something stronger is true. Let us examine the case of binary products in **Grp**, although all of the following can be said for any limits in any of the categories **Grp**, **Ab**, **Vect**$_k$, **Ring**, etc.

Take groups X_1 and X_2. We can form the product set $U(X_1) \times U(X_2)$, which comes equipped with projections

$$U(X_1) \xleftarrow{\ p_1\ } U(X_1) \times U(X_2) \xrightarrow{\ p_2\ } U(X_2).$$

I claim that there is exactly one group structure on the set $U(X_1) \times U(X_2)$ with the property that p_1 and p_2 are homomorphisms. To prove uniqueness, suppose that we have a group structure on $U(X_1) \times U(X_2)$ with this property. Take elements (x_1, x_2) and (x_1', x_2') of $U(X_1) \times U(X_2)$ and write $(x_1, x_2) \cdot (x_1', x_2') = (y_1, y_2)$. Since p_1 is a homomorphism,

$$y_1 = p_1(y_1, y_2) = p_1((x_1, x_2) \cdot (x_1', x_2')) = p_1(x_1, x_2) \cdot p_1(x_1', x_2') = x_1 \cdot x_1',$$

and similarly $y_2 = x_2 \cdot x_2'$. Hence

$$(x_1, x_2) \cdot (x_1', x_2') = (x_1 x_1', x_2 x_2').$$

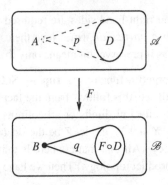

Figure 5.3 Creation of limits.

A similar argument shows that $(x_1, x_2)^{-1} = (x_1^{-1}, x_2^{-1})$ and that the identity element 1 of the group is $(1, 1)$. Now, for existence, define $\cdot$, $(\ )^{-1}$ and 1 by the formulas just given; it can then be checked that the group axioms are satisfied and that p_1 and p_2 are group homomorphisms. This proves the claim.

Write L for the set $U(X_1) \times U(X_2)$ equipped with this group structure. Then we have a cone

$$X_1 \xleftarrow{\ p_1\ } L \xrightarrow{\ p_2\ } X_2$$

in **Grp**. It is easy to check that this is, in fact, a *product* cone in **Grp**.

We can summarize this in language that is not tied to group theory. Given objects X_1 and X_2 of **Grp**,

- for any product cone on $(U(X_1), U(X_2))$ in **Set**, there is a unique cone on (X_1, X_2) in **Grp** whose image under U is the cone we started with;
- this cone on (X_1, X_2) is a product cone.

This suggests the following definition (Figure 5.3).

Definition 5.3.5   A functor $F \colon \mathscr{A} \to \mathscr{B}$ **creates limits (of shape I)** if whenever $D \colon \mathbf{I} \to \mathscr{A}$ is a diagram in $\mathscr{A}$,

- for any limit cone $\left(B \xrightarrow{\ q_I\ } FD(I)\right)_{I \in \mathbf{I}}$ on the diagram $F \circ D$, there is a unique cone $\left(A \xrightarrow{\ p_I\ } D(I)\right)_{I \in \mathbf{I}}$ on D such that $F(A) = B$ and $F(p_I) = q_I$ for all $I \in \mathbf{I}$;
- this cone $\left(A \xrightarrow{\ p_I\ } D(I)\right)_{I \in \mathbf{I}}$ is a limit cone on D.

The forgetful functors from **Grp**, **Ring**, ... to **Set** all create limits (Exercise 5.3.11). The word *creates* is explained by the following result.

Lemma 5.3.6 *Let $F: \mathscr{A} \to \mathscr{B}$ be a functor and $\mathbf{I}$ a small category. Suppose that $\mathscr{B}$ has, and F creates, limits of shape $\mathbf{I}$. Then $\mathscr{A}$ has, and F preserves, limits of shape $\mathbf{I}$.*

Proof Exercise 5.3.12. □

Since **Set** has all limits, it follows that all our categories of algebras have all limits, and that the forgetful functors preserve them.

Remark 5.3.7 There is something suspicious about Definition 5.3.5. It refers to *equality* of objects of a category, a relation that, as we saw on page 31, is usually too strict to be appropriate. It is almost always better to replace equality by isomorphism. If we replace equality by isomorphism throughout the definition of 'creates limits', we obtain a more healthy and inclusive notion. In the notation of Definition 5.3.5, we ask that if $F \circ D$ has a limit then there exists a cone on D whose image under F is a limit cone, and that every such cone is itself a limit cone.

In fact, what we are calling creation of limits should really be called *strict* creation of limits, with 'creation of limits' reserved for the more inclusive notion. That is how 'creates' is used in most of the literature. I have chosen to use the strict version here because it is slightly simpler to state, and because the examples at hand all satisfy the stricter condition.

Exercises

5.3.8 Taking the limit is a process that receives as its input a diagram in a category $\mathscr{A}$, and produces as its output a new object of $\mathscr{A}$. Later, we will see that this process is functorial (Proposition 6.1.4). Here you are asked to prove this in the case of binary products.

Let $\mathscr{A}$ be a category with binary products. Suppose that we have chosen for each pair (X, Y) of objects a product cone

$$X \xleftarrow{\ p_1^{X,Y}\ } X \times Y \xrightarrow{\ p_2^{X,Y}\ } Y.$$

Construct a functor $\mathscr{A} \times \mathscr{A} \to \mathscr{A}$ given on objects by $(X, Y) \mapsto X \times Y$.

5.3.9 Let $\mathscr{A}$ be a category with binary products. Prove directly that

$$\mathscr{A}(A, X \times Y) \cong \mathscr{A}(A, X) \times \mathscr{A}(A, Y)$$

naturally in $A, X, Y \in \mathscr{A}$. (This presupposes that we have chosen for each X and Y a product cone on (X, Y). By Exercise 5.3.8, the assignment $(X, Y) \mapsto X \times Y$ is then functorial, which it must be in order for 'naturally' to make sense.)

5.3.10   Prove that if a functor creates limits then it also reflects them.

5.3.11   It was shown in Example 5.3.4 that the forgetful functor $U: \mathbf{Grp} \to$ **Set** creates binary products.

(a) Using the formula for limits in **Set** (Example 5.1.22), prove that, in fact, U creates arbitrary limits.
(b) Satisfy yourself that the same is true if **Grp** is replaced by any other category of algebras such as **Ring, Ab** or $\mathbf{Vect}_k$.

5.3.12   Prove Lemma 5.3.6.

5.3.13   (a) An object P of a category $\mathscr{B}$ is **projective** if $\mathscr{B}(P, -): \mathscr{B} \to \mathbf{Set}$ preserves epics. (This means that if f is epic then so is $\mathscr{B}(P, f)$.) Let
$$\mathbf{Set} \xrightarrow[\;\;\;\;G\;\;\;\;]{\overset{F}{\underset{\perp}{\longrightarrow}}} \mathscr{B}$$
be an adjunction in which G preserves epics. Prove that $F(S)$ is projective for all sets S.
(b) Find a non-projective object of **Ab**.
(c) An object I of a category $\mathscr{B}$ is **injective** if it is projective in $\mathscr{B}^{\mathrm{op}}$, or equivalently if $\mathscr{B}(-, I): \mathscr{B}^{\mathrm{op}} \to \mathbf{Set}$ preserves epics. Show that all objects of $\mathbf{Vect}_k$ are injective, and find a non-injective object of **Ab**.

6

Adjoints, representables and limits

We have approached the idea of universal property from three different angles, producing three different formalisms: adjointness, representability, and limits. In this final chapter, we work out the connections between them.

In principle, anything that can be described in one of the three formalisms can also be described in the others. The situation is similar to that of cartesian and polar coordinates: anything that can be done in polar coordinates can in principle be done in cartesian coordinates, and vice versa, but some things are more gracefully done in one system than the other.

In comparing the three approaches, we will discover many of the fundamental results of category theory. Here are some highlights.

- Limits and colimits in functor categories work in the simplest possible way.
- The embedding of a category $\mathbf{A}$ into its presheaf category $[\mathbf{A}^{op}, \mathbf{Set}]$ preserves limits (but not colimits).
- The representables are the prime numbers of presheaves: every presheaf can be expressed canonically as a colimit of representables.
- A functor with a left adjoint preserves limits. Under suitable hypotheses, the converse holds too.
- Categories of presheaves $[\mathbf{A}^{op}, \mathbf{Set}]$ behave very much like the category of sets, the beginning of an incredible story that brings together the subjects of logic and geometry.

6.1 Limits in terms of representables and adjoints

There is more than one way to present the definition of limit. In Chapter 5, we used an explicit form of the definition that is particularly convenient for examples. But we will soon be developing the *theory* of limits and colimits, and for that, a rephrased form of the definition is useful. In fact, we rephrase it in two different ways: once in terms of representability, and once in terms of adjoints.

We begin by showing that cones are simply natural transformations of a special kind. To do this, we need some notation. Given categories $\mathbf{I}$ and $\mathscr{A}$ and an object $A \in \mathscr{A}$, there is a functor $\Delta A: \mathbf{I} \to \mathscr{A}$ with constant value A on objects and 1_A on maps. This defines, for each $\mathbf{I}$ and $\mathscr{A}$, the **diagonal functor**

$$\Delta: \mathscr{A} \to [\mathbf{I}, \mathscr{A}].$$

The name can be understood by considering the case in which $\mathbf{I}$ is the discrete category with two objects; then $[\mathbf{I}, \mathscr{A}] = \mathscr{A} \times \mathscr{A}$ and $\Delta(A) = (A, A)$.

Now, given a diagram $D: \mathbf{I} \to \mathscr{A}$ and an object $A \in \mathscr{A}$, a cone on D with vertex A is simply a natural transformation

$$\mathbf{I} \underset{D}{\overset{\Delta A}{\Longrightarrow}} \mathscr{A}.$$

Writing $\mathrm{Cone}(A, D)$ for the set of cones on D with vertex A, we therefore have

$$\mathrm{Cone}(A, D) = [\mathbf{I}, \mathscr{A}](\Delta A, D). \tag{6.1}$$

Thus, $\mathrm{Cone}(A, D)$ is functorial in A (contravariantly) and D (covariantly).

Here is our first rephrasing of the definition of limit.

Proposition 6.1.1 *Let $\mathbf{I}$ be a small category, $\mathscr{A}$ a category, and $D: \mathbf{I} \to \mathscr{A}$ a diagram. Then there is a one-to-one correspondence between limit cones on D and representations of the functor*

$$\mathrm{Cone}(-, D): \mathscr{A}^{\mathrm{op}} \to \mathbf{Set},$$

with the representing objects of $\mathrm{Cone}(-, D)$ *being the limit objects (that is, the vertices of the limit cones) of D.*

Briefly put: a limit of D is a representation of $[\mathbf{I}, \mathscr{A}](\Delta-, D)$.

Proof By Corollary 4.3.2, a representation of $\mathrm{Cone}(-, D)$ consists of a cone on D with a certain universal property. This is exactly the universal property in the definition of limit cone. $\square$

The proposition formalizes the thought that cones on a diagram D correspond one-to-one with maps into $\varprojlim_{\mathbf{I}} D$. It implies that if D has a limit then

$$\mathrm{Cone}(A, D) \cong \mathscr{A}\left(A, \varprojlim_{\mathbf{I}} D\right) \tag{6.2}$$

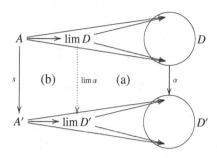

Figure 6.1 Illustration of Lemma 6.1.3.

naturally in A. The correspondence is given from left to right by

$$(f_I)_{I \in \mathbf{I}} \mapsto \bar{f}$$

(in the notation of Definition 5.1.19), and from right to left by

$$(p_I \circ g)_{I \in \mathbf{I}} \leftmapsto g$$

where $p_I \colon \varprojlim_{\mathbf{I}} D \to D(I)$ are the projections.

From Proposition 6.1.1 and Corollary 4.3.10 we deduce:

Corollary 6.1.2   *Limits are unique up to isomorphism.*   □

The characterization (6.1) of cones suggests that we might consider varying the diagram D as well as the vertex A. We are naturally led to ask questions such as: given a map $D \to D'$ between diagrams, is there an induced map between the limits of D and D'? The answer is yes (Figure 6.1):

Lemma 6.1.3   *Let* $\mathbf{I}$ *be a small category and* $\mathbf{I} \underset{D'}{\overset{D}{\Longrightarrow}} \Downarrow\alpha \; \mathscr{A}$ *a natural transformation. Let*

$$\left(\varprojlim_{\mathbf{I}} D \xrightarrow{\; p_I \;} D(I) \right)_{I \in \mathbf{I}} \quad and \quad \left(\varprojlim_{\mathbf{I}} D' \xrightarrow{\; p'_I \;} D'(I) \right)_{I \in \mathbf{I}}$$

be limit cones. Then:

(a) *there is a unique map* $\varprojlim_{\mathbf{I}} \alpha \colon \varprojlim_{\mathbf{I}} D \to \varprojlim_{\mathbf{I}} D'$ *such that for all* $I \in \mathbf{I}$, *the square*

$$\begin{array}{ccc} \varprojlim_{\mathbf{I}} D & \xrightarrow{\ p_I\ } & D(I) \\ {\scriptstyle \varprojlim_{\mathbf{I}} \alpha}\Big\downarrow & & \Big\downarrow{\scriptstyle \alpha_I} \\ \varprojlim_{\mathbf{I}} D' & \xrightarrow[\ p_I'\ ]{} & D'(I) \end{array}$$

commutes;

(b) *given cones* $\left(A \xrightarrow{f_I} D(I)\right)_{I \in \mathbf{I}}$ *and* $\left(A' \xrightarrow{f_I'} D'(I)\right)_{I \in \mathbf{I}}$ *and a map* $s \colon A \to A'$ *such that*

$$\begin{array}{ccc} A & \xrightarrow{\ f_I\ } & D(I) \\ {\scriptstyle s}\Big\downarrow & & \Big\downarrow{\scriptstyle \alpha_I} \\ A' & \xrightarrow[\ f_I'\ ]{} & D'(I) \end{array}$$

commutes for all $I \in \mathbf{I}$, *the square*

$$\begin{array}{ccc} A & \xrightarrow{\ \bar{f}\ } & \varprojlim_{\mathbf{I}} D \\ {\scriptstyle s}\Big\downarrow & & \Big\downarrow{\scriptstyle \varprojlim_{\mathbf{I}} \alpha} \\ A' & \xrightarrow[\ \bar{f'}\ ]{} & \varprojlim_{\mathbf{I}} D' \end{array}$$

also commutes.

Proof   Part (a) follows immediately from the fact that $\left(\varprojlim_{\mathbf{I}} D \xrightarrow{\alpha_I p_I} D'(I)\right)_{I \in \mathbf{I}}$ is a cone on D'. To prove (b), note that for each $I \in \mathbf{I}$, we have

$$p_I' \circ \left(\varprojlim_{\mathbf{I}} \alpha\right) \circ \bar{f} = \alpha_I \circ p_I \circ \bar{f} = \alpha_I \circ f_I = f_I' \circ s = p_I' \circ \overline{f'} \circ s.$$

So by Exercise 5.1.36(a), $\left(\varprojlim_{\mathbf{I}} \alpha\right) \circ \bar{f} = \overline{f'} \circ s$. $\square$

We can now give the second rephrasing of the definition of limit. It only applies when the category has *all* limits of the shape concerned.

Proposition 6.1.4   *Let* $\mathbf{I}$ *be a small category and* $\mathscr{A}$ *a category with all limits of shape* $\mathbf{I}$. *Then* $\varprojlim_{\mathbf{I}}$ *defines a functor* $[\mathbf{I}, \mathscr{A}] \to \mathscr{A}$, *and this functor is right adjoint to the diagonal functor.*

Proof   Choose for each $D \in [\mathbf{I}, \mathscr{A}]$ a limit cone on D, and call its vertex $\varprojlim_{\mathbf{I}} D$. For each map $\alpha \colon D \to D'$ in $[\mathbf{I}, \mathscr{A}]$, we have a canonical map $\varprojlim_{\mathbf{I}} \alpha \colon$

$\lim\limits_{\leftarrow I} D \to \lim\limits_{\leftarrow I} D'$, defined as in Lemma 6.1.3(a). This makes $\lim\limits_{\leftarrow I}$ into a functor. Proposition 6.1.1 implies that

$$[\mathbf{I}, \mathscr{A}](\Delta A, D) = \mathrm{Cone}(A, D) \cong \mathscr{A}\left(A, \lim\limits_{\leftarrow I} D\right)$$

naturally in A, and taking $s = 1_A$ in Lemma 6.1.3(b) tells us that the isomorphism is also natural in D.   □

To define the functor $\lim\limits_{\leftarrow I}$, we had to *choose* for each D a limit cone on D. This is a non-canonical choice. Nevertheless, different choices only affect the functor $\lim\limits_{\leftarrow I}$ up to natural isomorphism, by uniqueness of adjoints.

Exercises

6.1.5 Interpret all the theory of this section in the special case where $\mathbf{I}$ is the discrete category with two objects.

6.1.6 What is the content of Proposition 6.1.4 when $\mathbf{I}$ is a group and $\mathscr{A} = \mathbf{Set}$? What about the dual of Proposition 6.1.4?

6.2 Limits and colimits of presheaves

What do limits and colimits look like in functor categories $[\mathscr{A}, \mathscr{B}]$? In particular, what do they look like in presheaf categories $[\mathscr{A}^{\mathrm{op}}, \mathbf{Set}]$? More particularly still, what about limits and colimits of representables? Are they, too, representable?

We will answer all these questions. In order to do so, we first prove that representables preserve limits.

Representables preserve limits

Let us begin by recalling that, by definition of product, a map $A \to X \times Y$ amounts to a pair of maps $(A \to X, A \to Y)$. Here A, X and Y are objects of a category $\mathscr{A}$ with binary products. There is, therefore, a bijection

$$\mathscr{A}(A, X \times Y) \cong \mathscr{A}(A, X) \times \mathscr{A}(A, Y) \tag{6.3}$$

natural in $A, X, Y \in \mathscr{A}$.

Is this a special feature of products, or does some analogous statement hold for every kind of limit? Let us try equalizers. Suppose that $\mathscr{A}$ has equalizers, and write $\mathrm{Eq}\!\left(X \underset{t}{\overset{s}{\rightrightarrows}} Y\right)$ for the equalizer of maps s and t. By definition of equalizer, maps

$$A \;\to\; \mathrm{Eq}\!\left(X \underset{t}{\overset{s}{\rightrightarrows}} Y\right) \tag{6.4}$$

correspond one-to-one with maps $f\colon A \to X$ such that $s \circ f = t \circ f$. Now recall that s induces a map

$$s_* = \mathscr{A}(A, s)\colon\; \mathscr{A}(A, X) \to \mathscr{A}(A, Y),$$

and similarly for t. In this notation, what we have just said is that maps (6.4) correspond one-to-one with elements $f \in \mathscr{A}(A, X)$ such that

$$(\mathscr{A}(A, s))(f) = (\mathscr{A}(A, t))(f).$$

By the explicit formula for equalizers in **Set** (Example 5.1.12), such an f is exactly an element of the equalizer of $\mathscr{A}(A, s)$ and $\mathscr{A}(A, t)$. So, we have a canonical bijection

$$\mathscr{A}\!\left(A,\, \mathrm{Eq}\!\left(X \underset{t}{\overset{s}{\rightrightarrows}} Y\right)\right) \;\cong\; \mathrm{Eq}\!\left(\mathscr{A}(A, X) \underset{\mathscr{A}(A,t)}{\overset{\mathscr{A}(A,s)}{\rightrightarrows}} \mathscr{A}(A, Y)\right). \tag{6.5}$$

This looks something like our isomorphism (6.3) for products.

The isomorphisms (6.3) and (6.5) suggest that, more generally, we might have

$$\mathscr{A}\!\left(A, \lim_{\leftarrow \mathbf{I}} D\right) \;\cong\; \lim_{\leftarrow \mathbf{I}} \mathscr{A}(A, D) \tag{6.6}$$

naturally in $A \in \mathscr{A}$ and $D \in [\mathbf{I}, \mathscr{A}]$, whenever $\mathscr{A}$ is a category with limits of shape $\mathbf{I}$. Here $\mathscr{A}(A, D)$ is the functor

$$\begin{aligned} \mathscr{A}(A, D)\colon\quad \mathbf{I} \;&\to\; \mathbf{Set} \\ I \;&\mapsto\; \mathscr{A}(A, D(I)). \end{aligned}$$

This functor could also be written as $\mathscr{A}(A, D(-))$, and is the composite

$$\mathbf{I} \xrightarrow{\;D\;} \mathscr{A} \xrightarrow{\;\mathscr{A}(A,-)\;} \mathbf{Set}.$$

The conjectured isomorphism (6.6) states, essentially, that representables preserve limits. We now set about proving this.

Lemma 6.2.1 *Let* $\mathbf{I}$ *be a small category,* $\mathscr{A}$ *a locally small category,* $D\colon$ $\mathbf{I} \to \mathscr{A}$ *a diagram, and* $A \in \mathscr{A}$*. Then*

$$\mathrm{Cone}(A, D) \;\cong\; \lim_{\leftarrow \mathbf{I}} \mathscr{A}(A, D)$$

naturally in A and D.

Proof Like all functors from a small category into **Set**, the functor $\mathscr{A}(A, D)$ does have a limit, given by the explicit formula (5.16). According to this formula, $\varprojlim_{\mathbf{I}} \mathscr{A}(A, D)$ is the set of all families $(f_I)_{I \in \mathbf{I}}$ such that $f_I \in \mathscr{A}(A, D(I))$ for all $I \in \mathbf{I}$ and

$$(\mathscr{A}(A, Du))(f_I) = f_J \tag{6.7}$$

for all $I \xrightarrow{u} J$ in $\mathbf{I}$. But equation (6.7) just says that $(Du) \circ f_I = f_J$, so an element of $\varprojlim_{\mathbf{I}} \mathscr{A}(A, D)$ is nothing but a cone on D with vertex A. □

Proposition 6.2.2 (Representables preserve limits) *Let $\mathscr{A}$ be a locally small category and $A \in \mathscr{A}$. Then $\mathscr{A}(A, -) \colon \mathscr{A} \to$ **Set** preserves limits.*

Proof Let $\mathbf{I}$ be a small category and let $D \colon \mathbf{I} \to \mathscr{A}$ be a diagram that has a limit. Then

$$\mathscr{A}\left(A, \varprojlim_{\mathbf{I}} D\right) \cong \mathrm{Cone}(A, D) \cong \varprojlim_{\mathbf{I}} \mathscr{A}(A, D)$$

naturally in A. Here the first isomorphism is Proposition 6.1.1 (or more particularly, the isomorphism (6.2) that follows it), and the second is Lemma 6.2.1. □

Remark 6.2.3 Proposition 6.2.2 tells us that

$$\mathscr{A}\left(A, \varprojlim_{\mathbf{I}} D\right) \cong \varprojlim_{\mathbf{I}} \mathscr{A}(A, D). \tag{6.8}$$

To dualize Proposition 6.2.2, we replace $\mathscr{A}$ by $\mathscr{A}^{\mathrm{op}}$. Thus, $\mathscr{A}(-, A) \colon \mathscr{A}^{\mathrm{op}} \to$ **Set** preserves limits. A limit in $\mathscr{A}^{\mathrm{op}}$ is a colimit in $\mathscr{A}$, so $\mathscr{A}(-, A)$ transforms colimits in $\mathscr{A}$ into limits in **Set**:

$$\mathscr{A}\left(\varinjlim_{\mathbf{I}} D, A\right) \cong \varprojlim_{\mathbf{I}} \mathscr{A}(D, A). \tag{6.9}$$

The right-hand side is a *limit*, not a colimit! So even though (6.8) and (6.9) are dual statements, there are, in total, more limits than colimits involved. Somehow, limits have the upper hand.

For example, let X, Y and A be objects of a category $\mathscr{A}$, and suppose that the sum $X + Y$ exists. By definition of sum, a map $X + Y \to A$ amounts to a pair of maps $(X \to A, Y \to A)$. In other words, there is a canonical isomorphism

$$\mathscr{A}(X + Y, A) \cong \mathscr{A}(X, A) \times \mathscr{A}(Y, A).$$

This is the isomorphism (6.9) in the case where $\mathbf{I}$ is the discrete category with two objects.

Limits in functor categories

Earlier, we learned that it is sometimes useful to view functors as objects in their own right, rather than as maps of categories. For instance, when G is a group, functors $G \to$ **Set** are G-sets (Example 1.2.8), which one would usually regard as 'things' rather than 'maps'. This point of view leads to the concept of functor category.

We now begin an analysis of limits and colimits in functor categories $[\mathbf{A}, \mathscr{S}]$. Here $\mathbf{A}$ is small and $\mathscr{S}$ is locally small; these conditions together guarantee that $[\mathbf{A}, \mathscr{S}]$ is locally small. The most important cases for us will be $\mathscr{S} =$ **Set** and $\mathscr{S} =$ **Set**^{op}. For that reason, we will assume whenever necessary that $\mathscr{S}$ has all limits and colimits.

We show that limits and colimits in $[\mathbf{A}, \mathscr{S}]$ work in the simplest way imaginable. For instance, if $\mathscr{S}$ has binary products then so does $[\mathbf{A}, \mathscr{S}]$, and the product of two functors $X, Y \colon \mathbf{A} \to \mathscr{S}$ is the functor $X \times Y \colon \mathbf{A} \to \mathscr{S}$ given by

$$(X \times Y)(A) = X(A) \times Y(A)$$

for all $A \in \mathbf{A}$.

Notation 6.2.4 Let $\mathbf{A}$ and $\mathscr{S}$ be categories. For each $A \in \mathbf{A}$, there is a functor

$$\mathrm{ev}_A \colon \quad [\mathbf{A}, \mathscr{S}] \quad \to \quad \mathscr{S}$$
$$X \quad \mapsto \quad X(A),$$

called **evaluation** at A. We will be working with diagrams in $[\mathbf{A}, \mathscr{S}]$, and given such a diagram $D \colon \mathbf{I} \to [\mathbf{A}, \mathscr{S}]$, we have for each $A \in \mathbf{A}$ a functor

$$\mathrm{ev}_A \circ D \colon \quad \mathbf{I} \quad \to \quad \mathscr{S}$$
$$I \quad \mapsto \quad D(I)(A).$$

We write $\mathrm{ev}_A \circ D$ as $D(-)(A)$.

Theorem 6.2.5 (Limits in functor categories) *Let* $\mathbf{A}$ *and* $\mathbf{I}$ *be small categories and* $\mathscr{S}$ *a locally small category. Let* $D \colon \mathbf{I} \to [\mathbf{A}, \mathscr{S}]$ *be a diagram, and suppose that for each* $A \in \mathbf{A}$, *the diagram* $D(-)(A) \colon \mathbf{I} \to \mathscr{S}$ *has a limit. Then there is a cone on* D *whose image under* ev_A *is a limit cone on* $D(-)(A)$ *for each* $A \in \mathbf{A}$. *Moreover, any such cone on* D *is a limit cone.*

Theorem 6.2.5 is often expressed as a slogan:

Limits in a functor category are computed pointwise.

The 'points' in the word 'pointwise' are the objects of $\mathbf{A}$. The slogan means, for example, that given two functors $X, Y \in [\mathbf{A}, \mathscr{S}]$, their product can be computed by first taking the product $X(A) \times Y(A)$ in $\mathscr{S}$ for each 'point' A, then assembling them to form a functor $X \times Y$.

Of course, Theorem 6.2.5 has a dual, stating that colimits in a functor category are also computed pointwise.

Proof of Theorem 6.2.5   Take for each $A \in \mathbf{A}$ a limit cone

$$\left(L(A) \xrightarrow{p_{I,A}} D(I)(A) \right)_{I \in \mathbf{I}} \tag{6.10}$$

on the diagram $D(-)(A) \colon \mathbf{I} \to \mathscr{S}$. We prove two statements:

(a) there is exactly one way of extending L to a functor on $\mathbf{A}$ with the property that $\left(L \xrightarrow{p_I} D(I) \right)_{I \in \mathbf{I}}$ is a cone on D;

(b) this cone $\left(L \xrightarrow{p_I} D(I) \right)_{I \in \mathbf{I}}$ is a limit cone.

The theorem will follow immediately.

For (a), take a map $f \colon A \to A'$ in $\mathbf{A}$. Lemma 6.1.3(a) applied to the natural transformation

$$\mathbf{I} \overset{\overset{D(-)(A)}{\displaystyle\longrightarrow}}{\underset{\underset{D(-)(A')}{\displaystyle\longrightarrow}}{\Downarrow D(-)(f)}} \mathscr{S}$$

implies that there is a unique map $L(f) \colon L(A) \to L(A')$ such that for all $I \in \mathbf{I}$, the square

$$\begin{array}{ccc} L(A) & \xrightarrow{\;\;p_{I,A}\;\;} & D(I)(A) \\ {\scriptstyle L(f)}\big\downarrow & & \big\downarrow{\scriptstyle D(I)(f)} \\ L(A') & \xrightarrow[\;\;p_{I,A'}\;\;]{} & D(I)(A') \end{array} \tag{6.11}$$

commutes. (This is our *definition* of $L(f)$.) We have now defined L on objects and maps of $\mathbf{A}$. It is easy to check that L preserves composition and identities, and is therefore a functor $L \colon \mathbf{A} \to \mathscr{S}$. Moreover, the commutativity of diagram (6.11) says exactly that for each $I \in \mathbf{I}$, the family $\left(L(A) \xrightarrow{p_{I,A}} D(I)(A) \right)_{A \in \mathbf{A}}$ is a natural transformation

$$\mathbf{A} \overset{\overset{L}{\displaystyle\longrightarrow}}{\underset{\underset{D(I)}{\displaystyle\longrightarrow}}{\Downarrow p_I}} \mathscr{S}.$$

So we have a family $\left(L \xrightarrow{p_I} D(I) \right)_{I \in \mathbf{I}}$ of maps in $[\mathbf{A}, \mathscr{S}]$, and from the fact that (6.10) is a cone on $D(-)(A)$ for each $A \in \mathbf{A}$, it follows immediately that $\left(L \xrightarrow{p_I} D(I) \right)_{I \in \mathbf{I}}$ is a cone on D.

For (b), let $X \in [\mathbf{A}, \mathscr{S}]$ and let $\left(X \xrightarrow{q_I} D(I)\right)_{I \in \mathbf{I}}$ be a cone on D in $[\mathbf{A}, \mathscr{S}]$. For each $A \in \mathbf{A}$, we have a cone

$$\left(X(A) \xrightarrow{q_{I,A}} D(I)(A)\right)_{I \in \mathbf{I}}$$

on $D(-)(A)$ in $\mathscr{S}$, so there is a unique map $\bar{q}_A \colon X(A) \to L(A)$ such that $p_{I,A} \circ \bar{q}_A = q_{I,A}$ for all $I \in \mathbf{I}$. It only remains to prove that $\bar{q}_A$ is natural in A, and that follows from Lemma 6.1.3(b).   □

Theorem 6.2.5 has many important consequences. We begin by recording a cruder form of the theorem (and its dual), which we will use repeatedly.

Corollary 6.2.6 *Let* $\mathbf{I}$ *and* $\mathbf{A}$ *be small categories, and* $\mathscr{S}$ *a locally small category. If* $\mathscr{S}$ *has all limits (respectively, colimits) of shape* $\mathbf{I}$ *then so does* $[\mathbf{A}, \mathscr{S}]$, *and for each* $A \in \mathbf{A}$, *the evaluation functor* $\mathrm{ev}_A \colon [\mathbf{A}, \mathscr{S}] \to \mathscr{S}$ *preserves them.*   □

Warning 6.2.7 If $\mathscr{S}$ does *not* have all limits of shape $\mathbf{I}$ then $[\mathbf{A}, \mathscr{S}]$ may contain limits of shape $\mathbf{I}$ that are not computed pointwise, that is, are not preserved by all the evaluation functors. Examples can be constructed, as in Section 3.3 of Kelly (1982).

Theorem 6.2.5 will also help us to prove that limits commute with limits, in the following sense. Take categories $\mathbf{I}$, $\mathbf{J}$ and $\mathscr{S}$. There are isomorphisms of categories

$$[\mathbf{I}, [\mathbf{J}, \mathscr{S}]] \cong [\mathbf{I} \times \mathbf{J}, \mathscr{S}] \cong [\mathbf{J}, [\mathbf{I}, \mathscr{S}]].$$

(See Remark 4.1.23(c) and Exercise 1.2.25.) Under these isomorphisms, a functor $D \colon \mathbf{I} \times \mathbf{J} \to \mathscr{S}$ corresponds to the functors

$$\begin{array}{ccccccccc} D^{\bullet} \colon & \mathbf{I} & \to & [\mathbf{J}, \mathscr{S}] & \quad \text{and} \quad & D_{\bullet} \colon & \mathbf{J} & \to & [\mathbf{I}, \mathscr{S}] \\ & I & \mapsto & D(I, -) & & & J & \mapsto & D(-, J). \end{array}$$

Supposing that $\mathscr{S}$ has all limits, so do the various functor categories, by Corollary 6.2.6. In particular, there is an object $\lim_{\leftarrow \mathbf{I}} D^{\bullet}$ of $[\mathbf{J}, \mathscr{S}]$. This is itself a diagram in $\mathscr{S}$, so we obtain in turn an object $\lim_{\leftarrow \mathbf{J}} \lim_{\leftarrow \mathbf{I}} D^{\bullet}$ of $\mathscr{S}$. Alternatively, we can take limits in the other order, producing an object $\lim_{\leftarrow \mathbf{I}} \lim_{\leftarrow \mathbf{J}} D_{\bullet}$ of $\mathscr{S}$. And there is a third possibility: taking the limit of D itself, we obtain another object $\lim_{\leftarrow \mathbf{I} \times \mathbf{J}} D$ of $\mathscr{S}$. The next result states that these three objects are the same. That is, it makes no difference what order we take limits in.

Proposition 6.2.8 (Limits commute with limits) *Let* $\mathbf{I}$ *and* $\mathbf{J}$ *be small categories. Let* $\mathscr{S}$ *be a locally small category with limits of shape* $\mathbf{I}$ *and of shape*

J. *Then for all* $D: \mathbf{I} \times \mathbf{J} \to \mathscr{S}$, *we have*

$$\varprojlim_{\leftarrow \mathbf{J}} \varprojlim_{\leftarrow \mathbf{I}} D^{\bullet} \cong \varprojlim_{\leftarrow \mathbf{I} \times \mathbf{J}} D \cong \varprojlim_{\leftarrow \mathbf{I}} \varprojlim_{\leftarrow \mathbf{J}} D_{\bullet},$$

and all these limits exist. In particular, $\mathscr{S}$ *has limits of shape* $\mathbf{I} \times \mathbf{J}$.

This is sometimes half-jokingly called Fubini's theorem, as it is something like changing the order of integration in a double integral. The analogy is more appealing with *co*limits, since, like integrals, colimits can be thought of as a context-sensitive version of sums.

Proof By symmetry, it is enough to prove the first isomorphism. Since $\mathscr{S}$ has limits of shape $\mathbf{I}$, so does $[\mathbf{J}, \mathscr{S}]$ (by Corollary 6.2.6). So $\varprojlim_{\leftarrow \mathbf{I}} D^{\bullet}$ exists; it is an object of $[\mathbf{J}, \mathscr{S}]$. Since $\mathscr{S}$ has limits of shape $\mathbf{J}$, $\varprojlim_{\leftarrow \mathbf{J}} \varprojlim_{\leftarrow \mathbf{I}} D^{\bullet}$ exists; it is an object of $\mathscr{S}$. Then for $S \in \mathscr{S}$,

$$\mathscr{S}\left(S, \varprojlim_{\leftarrow \mathbf{J}} \varprojlim_{\leftarrow \mathbf{I}} D^{\bullet}\right) \cong [\mathbf{J}, \mathscr{S}]\left(\Delta S, \varprojlim_{\leftarrow \mathbf{I}} D^{\bullet}\right)$$

$$\cong [\mathbf{I}, [\mathbf{J}, \mathscr{S}]](\Delta(\Delta S), D^{\bullet})$$

$$\cong [\mathbf{I} \times \mathbf{J}, \mathscr{S}](\Delta S, D)$$

naturally in S. The first two steps each follow from Proposition 6.1.1. The third uses the isomorphism $[\mathbf{I}, [\mathbf{J}, \mathscr{S}]] \cong [\mathbf{I} \times \mathbf{J}, \mathscr{S}]$, under which $\Delta(\Delta S)$ corresponds to ΔS and $D^{\bullet}$ corresponds to D.

Hence $\varprojlim_{\leftarrow \mathbf{J}} \varprojlim_{\leftarrow \mathbf{I}} D^{\bullet}$ is a representing object for the functor $[\mathbf{I} \times \mathbf{J}, \mathscr{S}](\Delta -, D)$. By Proposition 6.1.1 again, this says that $\varprojlim_{\leftarrow \mathbf{I} \times \mathbf{J}} D$ exists and is isomorphic to $\varprojlim_{\leftarrow \mathbf{J}} \varprojlim_{\leftarrow \mathbf{I}} D^{\bullet}$. □

Example 6.2.9 When $\mathbf{I} = \mathbf{J} = \boxed{\bullet \qquad \bullet}$, Proposition 6.2.8 says that binary products commute with binary products: if $\mathscr{S}$ has binary products and $S_{11}, S_{12}, S_{21}, S_{22} \in \mathscr{S}$ then the 4-fold product $\prod_{i,j \in \{1,2\}} S_{ij}$ exists and satisfies

$$(S_{11} \times S_{21}) \times (S_{12} \times S_{22}) \cong \prod_{i,j \in \{1,2\}} S_{ij} \cong (S_{11} \times S_{12}) \times (S_{21} \times S_{22}).$$

More generally, it makes no difference what order we write products in or where we put the brackets: there are canonical isomorphisms

$$S \times T \cong T \times S,$$

$$(S \times T) \times U \cong S \times (T \times U)$$

in any category with binary products. If there is also a terminal object 1, there are further canonical isomorphisms

$$S \times 1 \cong S \cong 1 \times S.$$

Warning 6.2.10   The dual of Proposition 6.2.8 states that colimits commute with colimits. For instance,

$$(S_{11} + S_{21}) + (S_{12} + S_{22}) \cong (S_{11} + S_{12}) + (S_{21} + S_{22})$$

in any category $\mathscr{S}$ with binary sums. But limits do *not* in general commute with colimits. For instance, in general,

$$(S_{11} + S_{21}) \times (S_{12} + S_{22}) \not\cong (S_{11} \times S_{12}) + (S_{21} \times S_{22}).$$

A counterexample is given by taking $\mathscr{S}$ = **Set** and each S_{ij} to be a one-element set. Then the left-hand side has $(1 + 1) \times (1 + 1) = 4$ elements, whereas the right-hand side has $(1 \times 1) + (1 \times 1) = 2$ elements.

Here are two further consequences of Theorem 6.2.5.

Corollary 6.2.11   *Let* **A** *be a small category. Then* $[\mathbf{A}^{\mathrm{op}}, \mathbf{Set}]$ *has all limits and colimits, and for each* $A \in \mathbf{A}$*, the evaluation functor* $\mathrm{ev}_A : [\mathbf{A}^{\mathrm{op}}, \mathbf{Set}] \to$ **Set** *preserves them.*

Proof   Since **Set** has all limits and colimits, this is immediate from Corollary 6.2.6. $\square$

Corollary 6.2.12   *The Yoneda embedding* $H_{\bullet} : \mathbf{A} \to [\mathbf{A}^{\mathrm{op}}, \mathbf{Set}]$ *preserves limits, for any small category* **A**.

Proof   Let $D : \mathbf{I} \to \mathbf{A}$ be a diagram in **A**, and let $\left(\lim\limits_{\leftarrow \mathbf{I}} D \xrightarrow{p_I} D(I)\right)_{I \in \mathbf{I}}$ be a limit cone. For each $A \in \mathbf{A}$, the composite functor

$$\mathbf{A} \xrightarrow{H_{\bullet}} [\mathbf{A}^{\mathrm{op}}, \mathbf{Set}] \xrightarrow{\mathrm{ev}_A} \mathbf{Set}$$

is H^A, which preserves limits (Proposition 6.2.2). So for each $A \in \mathbf{A}$,

$$\left(\mathrm{ev}_A \, H_{\bullet}\!\left(\lim_{\leftarrow \mathbf{I}} D\right) \xrightarrow{\mathrm{ev}_A \, H_{\bullet}(p_I)} \mathrm{ev}_A \, H_{\bullet}(D(I))\right)_{I \in \mathbf{I}}$$

is a limit cone. But then, by the 'moreover' part of Theorem 6.2.5 applied to the diagram $H_{\bullet} \circ D$ in $[\mathbf{A}^{\mathrm{op}}, \mathbf{Set}]$, the cone

$$\left(H_{\bullet}\!\left(\lim_{\leftarrow \mathbf{I}} D\right) \xrightarrow{H_{\bullet}(p_I)} H_{\bullet}(D(I))\right)_{I \in \mathbf{I}}$$

is also a limit, as required. $\square$

Example 6.2.13   Let **A** be a category with binary products. Corollary 6.2.12 implies that for all $X, Y \in \mathbf{A}$,

$$H_{X \times Y} \cong H_X \times H_Y \tag{6.12}$$

in $[\mathbf{A}^{\mathrm{op}}, \mathbf{Set}]$. When evaluated at a particular object A, this says that

$$\mathbf{A}(A, X \times Y) \cong \mathbf{A}(A, X) \times \mathbf{A}(A, Y)$$

(using the fact that products are computed pointwise). This is the isomorphism (6.3) that we met at the beginning of this section.

Suppose that we view $\mathbf{A}$ as a subcategory of $[\mathbf{A}^{\mathrm{op}}, \mathbf{Set}]$, identifying $A \in \mathbf{A}$ with the representable $H_A \in [\mathbf{A}^{\mathrm{op}}, \mathbf{Set}]$ as in Figure 4.1. Then the isomorphism (6.12) means that given two objects of $\mathbf{A}$ whose product we want to form, it makes no difference whether we think of the product as taking place in $\mathbf{A}$ or $[\mathbf{A}^{\mathrm{op}}, \mathbf{Set}]$. Similarly, if $\mathbf{A}$ has all limits, taking limits does not help us to escape from $\mathbf{A}$ into the rest of $[\mathbf{A}^{\mathrm{op}}, \mathbf{Set}]$: any limit of representable presheaves is again representable.

Warning 6.2.14 The Yoneda embedding does *not* preserve colimits. For example, if $\mathbf{A}$ has an initial object 0 then H_0 is not initial, since $H_0(0) = \mathbf{A}(0, 0)$ is a one-element set, whereas the initial object of $[\mathbf{A}^{\mathrm{op}}, \mathbf{Set}]$ is the presheaf with constant value $\emptyset$. We investigate colimits of representables next.

Every presheaf is a colimit of representables

We now know that the Yoneda embedding preserves limits but not colimits. In fact, the situation for colimits is at the opposite extreme from the situation for limits: by taking colimits of representable presheaves, we can obtain any presheaf we like! This is the last main result of this section.

Every positive integer can be expressed as a product of primes in an essentially unique way. Somewhat similarly, every presheaf can be expressed as a colimit of representables in a canonical (though not unique) way. The representables are the building blocks of presheaves.

For a different analogy, recall that any complex function holomorphic in a neighbourhood of 0 has a power series expansion, such as

$$e^z = 1 + z + \frac{z^2}{2!} + \frac{z^3}{3!} + \cdots .$$

In this sense, the power functions $z \mapsto z^n$ are the building blocks of holomorphic functions. We could even take the analogy further: $(\ )^n$ is like a representable $\mathrm{Hom}(n, -)$, and in the categorical context, quotients and sums are types of colimit.

Before we state and prove the theorem, let us look at an easy special case.

Example 6.2.15 Let $\mathbf{A}$ be the discrete category with two objects, K and L. A presheaf X on $\mathbf{A}$ is just a pair $(X(K), X(L))$ of sets, and $[\mathbf{A}^{\mathrm{op}}, \mathbf{Set}] \cong \mathbf{Set} \times \mathbf{Set}$. There are two representables, H_K and H_L, given by

$$H_A(B) = \mathbf{A}(B, A) \cong \begin{cases} 1 & \text{if } A = B, \\ \emptyset & \text{if } A \neq B \end{cases}$$

$(A, B \in \{K, L\})$. Identifying $[\mathbf{A}^{\mathrm{op}}, \mathbf{Set}]$ with $\mathbf{Set} \times \mathbf{Set}$, we have $H_K \cong (1, \emptyset)$ and $H_L \cong (\emptyset, 1)$. Every object of $\mathbf{Set} \times \mathbf{Set}$ is a sum of copies of $(1, \emptyset)$ and $(\emptyset, 1)$. Suppose, for instance, that $X(K)$ has three elements and $X(L)$ has two elements. Then

$$(X(K), X(L)) \cong (1, \emptyset) + (1, \emptyset) + (1, \emptyset) + (\emptyset, 1) + (\emptyset, 1)$$

in $\mathbf{Set} \times \mathbf{Set}$. Equivalently,

$$X \cong H_K + H_K + H_K + H_L + H_L$$

in $[\mathbf{A}^{\mathrm{op}}, \mathbf{Set}]$, exhibiting X as a sum of representables.

In this example, X is expressed as a sum of five representables, that is, a sum indexed by the set $X(K) + X(L)$ of 'elements' of X. A sum is a colimit over a discrete category. In the general case, a presheaf X on a category $\mathbf{A}$ is expressed as a colimit over a category whose objects can be thought of as the 'elements' of X. This is made precise by the following definition.

Definition 6.2.16 Let $\mathbf{A}$ be a category and X a presheaf on $\mathbf{A}$. The **category of elements** $\mathbf{E}(X)$ of X is the category in which:

- objects are pairs (A, x) with $A \in \mathbf{A}$ and $x \in X(A)$;
- maps $(A', x') \to (A, x)$ are maps $f \colon A' \to A$ in $\mathbf{A}$ such that $(Xf)(x) = x'$.

There is a projection functor $P \colon \mathbf{E}(X) \to \mathbf{A}$ defined by $P(A, x) = A$ and $P(f) = f$.

The following 'density theorem' states that every presheaf is a colimit of representables in a canonical way. It is secretly dual to the Yoneda lemma. This becomes apparent if one expresses both in suitably lofty categorical language (that of ends, or that of bimodules); but that is beyond the scope of this book.

Theorem 6.2.17 (Density) *Let $\mathbf{A}$ be a small category and X a presheaf on $\mathbf{A}$. Then X is the colimit of the diagram*

$$\mathbf{E}(X) \xrightarrow{P} \mathbf{A} \xrightarrow{H_\bullet} [\mathbf{A}^{\mathrm{op}}, \mathbf{Set}]$$

in $[\mathbf{A}^{\mathrm{op}}, \mathbf{Set}]$; that is, $X \cong \varinjlim_{\to \mathbf{I}} (H_\bullet \circ P)$.

Proof First note that since **A** is small, so too is $\mathbf{E}(X)$. Hence $H_\bullet \circ P$ really is a diagram in our customary sense (Definition 5.1.18).

Now let $Y \in [\mathbf{A}^{\mathrm{op}}, \mathbf{Set}]$. A cocone on $H_\bullet \circ P$ with vertex Y is a family

$$\left(H_A \xrightarrow{\alpha_{A,x}} Y \right)_{A \in \mathbf{A}, x \in X(A)}$$

of natural transformations with the property that for all maps $A' \xrightarrow{f} A$ in **A** and all $x \in X(A)$, the diagram

$$\begin{array}{c} H_{A'} \\ {\scriptstyle H_f} \downarrow \quad\searrow^{\alpha_{A',(Xf)(x)}} \\ \qquad\qquad\nearrow \quad Y \\ H_A \quad {\scriptstyle \alpha_{A,x}} \end{array}$$

commutes.

Equivalently (by the Yoneda lemma), a cocone on $H_\bullet \circ P$ with vertex Y is a family

$$(y_{A,x})_{A \in \mathbf{A}, x \in X(A)},$$

with $y_{A,x} \in Y(A)$, such that for all maps $A' \xrightarrow{f} A$ in **A** and all $x \in X(A)$,

$$(Yf)(y_{A,x}) = y_{A',(Xf)(x)}.$$

To see this, note that if $\alpha_{A,x} \in [\mathbf{A}^{\mathrm{op}}, \mathbf{Set}](H_A, Y)$ corresponds to $y_{A,x} \in Y(A)$, then $\alpha_{A,x} \circ H_f \in [\mathbf{A}^{\mathrm{op}}, \mathbf{Set}](H_{A'}, Y)$ corresponds to $(Yf)(y_{A,x}) \in Y(A')$.

Equivalently (writing $y_{A,x}$ as $\bar{\alpha}_A(x)$), it is a family

$$\left(X(A) \xrightarrow{\bar{\alpha}_A} Y(A) \right)_{A \in \mathbf{A}}$$

of functions with the property that for all maps $A' \xrightarrow{f} A$ in **A** and all $x \in X(A)$,

$$(Yf)(\bar{\alpha}_A(x)) = \bar{\alpha}_{A'}((Xf)(x)).$$

But this is simply a natural transformation $\bar{\alpha} \colon X \to Y$. So we have, for each $Y \in [\mathbf{A}^{\mathrm{op}}, \mathbf{Set}]$, a canonical bijection

$$[\mathbf{E}(X), [\mathbf{A}^{\mathrm{op}}, \mathbf{Set}]](H_\bullet \circ P, \Delta Y) \cong [\mathbf{A}^{\mathrm{op}}, \mathbf{Set}](X, Y).$$

Hence X is the colimit of $H_\bullet \circ P$. $\square$

Example 6.2.18 In Example 6.2.15, we expressed a particular presheaf X as a sum of representables. Let us check that the way we did this is a special case of the general construction in the density theorem.

Since **A** is discrete, the category of elements $\mathbf{E}(X)$ is also discrete; it is the set $X(K) + X(L)$ with five elements. The projection $P \colon \mathbf{E}(X) \to \mathbf{A}$ sends three of the elements to K and the other two to L, so the diagram $H_\bullet \circ P \colon E(X) \to [\mathbf{A}^{op}, \mathbf{Set}]$ sends three of the elements to H_K and two to H_L. The colimit of $H_\bullet \circ P$ is the sum of these five representables, which is X, just as in Example 6.2.15.

Remarks 6.2.19 (a) The term 'category of elements' is compatible with the generalized element terminology introduced in Definition 4.1.25. A generalized element of an object X is just a map into X, say $Z \to X$; but, as explained after that definition, we often focus on certain special shapes Z. Now suppose that we are working in a presheaf category $[\mathbf{A}^{op}, \mathbf{Set}]$. Among all presheaves, the representables have a special status, so we might be especially interested in generalized elements of representable shape. The Yoneda lemma implies that for a presheaf X, the generalized elements of X of representable shape correspond to pairs (A, x) with $A \in \mathbf{A}$ and $x \in X(A)$. In other words, they are the objects of the category of elements.

(b) In topology, a subspace A of a space B is called dense if every point in B can be obtained as a limit of points in A. This provides some explanation for the name of Theorem 6.2.17: the category $\mathbf{A}$ is 'dense' in $[\mathbf{A}^{op}, \mathbf{Set}]$ because every object of $[\mathbf{A}^{op}, \mathbf{Set}]$ can be obtained as a colimit of objects of $\mathbf{A}$.

Exercises

6.2.20 Fix a small category $\mathbf{A}$.

(a) Let $\mathscr{S}$ be a locally small category with pullbacks. Show that a natural transformation

$$\mathbf{A} \underset{Y}{\overset{X}{\rightrightarrows}} {\Downarrow \alpha}\ \mathscr{S}$$

is monic (as a map in $[\mathbf{A}, \mathscr{S}]$) if and only if α_A is monic for all $A \in \mathbf{A}$. (Hint: use Lemma 5.1.32.)

(b) Describe explicitly the monics and epics in $[\mathbf{A}^{op}, \mathbf{Set}]$.

(c) Can you do part (b) without relying on the fact that limits and colimits of presheaves are computed pointwise?

6.2.21 (a) Prove that representables have the following connectedness property: given a locally small category $\mathscr{A}$ and $A \in \mathscr{A}$, if $X, Y \in [\mathscr{A}^{op}, \mathbf{Set}]$ with $H_A \cong X + Y$, then either X or Y is the constant functor $\emptyset$.

(b) Deduce that the sum of two representables is never representable.

6.2.22 Show how a category of elements can be described as a comma category.

6.2.23 Let X be a presheaf on a locally small category. Show that X is representable if and only if its category of elements has a terminal object.

(Since a terminal object is a limit of the empty diagram, this implies that the concept of representability can be derived from the concept of limit. Since a terminal object of a category $\mathscr{E}$ is also a right adjoint to the unique functor $\mathscr{E} \to \mathbf{1}$, the concept of representability can also be derived from the concept of adjoint.)

6.2.24 Prove that every slice of a presheaf category is again a presheaf category. That is, given a small category $\mathbf{A}$ and a presheaf X on $\mathbf{A}$, prove that $[\mathbf{A}^{\mathrm{op}}, \mathbf{Set}]/X$ is equivalent to $[\mathbf{B}^{\mathrm{op}}, \mathbf{Set}]$ for some small category $\mathbf{B}$.

6.2.25 Let $F \colon \mathbf{A} \to \mathbf{B}$ be a functor between small categories. For each object $B \in \mathbf{B}$, there is a comma category $(F \Rightarrow B)$ (defined dually to the comma category in Example 2.3.4), and there is a projection functor $P_B \colon (F \Rightarrow B) \to \mathbf{A}$.

(a) Let $X \colon \mathbf{A} \to \mathscr{S}$ be a functor from $\mathbf{A}$ to a category $\mathscr{S}$ with small colimits. For each $B \in \mathbf{B}$, let $(\mathrm{Lan}_F X)(B)$ be the colimit of the diagram

$$(F \Rightarrow B) \xrightarrow{P_B} \mathbf{A} \xrightarrow{X} \mathscr{S}.$$

Show that this defines a functor $\mathrm{Lan}_F X \colon \mathbf{B} \to \mathscr{S}$, and that for functors $Y \colon \mathbf{B} \to \mathscr{S}$, there is a canonical bijection between natural transformations $\mathrm{Lan}_F X \to Y$ and natural transformations $X \to Y \circ F$.

(b) Deduce that for any category $\mathscr{S}$ with small colimits, the functor

$$- \circ F \colon [\mathbf{B}, \mathscr{S}] \to [\mathbf{A}, \mathscr{S}]$$

has a left adjoint. (This left adjoint, Lan_F, is called **left Kan extension** along F.)

(c) Part (b) and its dual imply that when $\mathscr{S}$ has small limits and colimits, the functor $- \circ F$ has both left and right adjoints. Revisit Exercise 2.1.16 with this in mind, taking F to be either the unique functor $\mathbf{1} \to G$ or the unique functor $G \to \mathbf{1}$.

6.3 Interactions between adjoint functors and limits

We saw in Proposition 4.1.11 that any set-valued functor with a left adjoint is representable, and in Proposition 6.2.2 that any representable preserves limits.

Hence, any set-valued functor with a left adjoint preserves limits. In fact, this conclusion holds not only for set-valued functors, but in complete generality.

Theorem 6.3.1   *Let $\mathscr{A} \underset{G}{\overset{F}{\underset{\perp}{\rightleftarrows}}} \mathscr{B}$ be an adjunction. Then F preserves colimits and G preserves limits.*

Proof   By duality, it is enough to prove that G preserves limits. Let $D: \mathbf{I} \to \mathscr{B}$ be a diagram for which a limit exists. Then

$$\mathscr{A}\Big(A, G\big(\varprojlim_{\mathbf{I}} D\big)\Big) \cong \mathscr{B}\Big(F(A), \varprojlim_{\mathbf{I}} D\Big) \qquad (6.13)$$

$$\cong \varprojlim_{\mathbf{I}} \mathscr{B}(F(A), D) \qquad (6.14)$$

$$\cong \varprojlim_{\mathbf{I}} \mathscr{A}(A, G \circ D) \qquad (6.15)$$

$$\cong \mathrm{Cone}(A, G \circ D) \qquad (6.16)$$

naturally in $A \in \mathscr{A}$. Here, the isomorphism (6.13) is by adjointness, (6.14) is because representables preserve limits, (6.15) is by adjointness again, and (6.16) is by Lemma 6.2.1. So $G\big(\varprojlim_{\mathbf{I}} D\big)$ represents $\mathrm{Cone}(-, G \circ D)$; that is, it is a limit of $G \circ D$.   □

Example 6.3.2   Forgetful functors from categories of algebras to **Set** have left adjoints, but hardly ever right adjoints. Correspondingly, they preserve all limits, but rarely all colimits.

Example 6.3.3   Every set B gives rise to an adjunction $(- \times B) \dashv (-)^B$ of functors from **Set** to **Set** (Example 2.1.6). So $- \times B$ preserves colimits and $(-)^B$ preserves limits. In particular, $- \times B$ preserves finite sums and $(-)^B$ preserves finite products, giving isomorphisms

$$0 \times B \cong 0, \qquad (A_1 + A_2) \times B \cong (A_1 \times B) + (A_2 \times B), \qquad (6.17)$$
$$1^B \cong 1, \qquad (A_1 \times A_2)^B \cong A_1^B \times A_2^B. \qquad (6.18)$$

These are the analogues of standard rules of arithmetic. (See also Example 6.2.9 and the 'Digression on arithmetic' on page 69.) Indeed, if we know (6.17) and (6.18) for just finite sets then by taking cardinality on both sides, we obtain exactly these standard rules. The natural numbers are, after all, just the isomorphism classes of finite sets.

Example 6.3.4   Given a category $\mathscr{A}$ with all limits of shape $\mathbf{I}$, we have the adjunction $\mathscr{A} \underset{\varprojlim_{\mathbf{I}}}{\overset{\Delta}{\underset{\perp}{\rightleftarrows}}} [\mathbf{I}, \mathscr{A}]$ (Proposition 6.1.4). Hence $\varprojlim_{\mathbf{I}}$ preserves limits, or equivalently, limits of shape **I** commute with (all) limits. This gives another proof that limits commute with limits (Proposition 6.2.8), at least in the case where the category has all limits of one of the shapes concerned.

Example 6.3.5 Theorem 6.3.1 is often used to prove that a functor does *not* have an adjoint. For instance, it was claimed in Example 2.1.3(e) that the forgetful functor U: **Field** $\to$ **Set** does not have a left adjoint. We can now prove this. If U had a left adjoint F: **Set** $\to$ **Field**, then F would preserve colimits, and in particular, initial objects. Hence $F(\emptyset)$ would be an initial object of **Field**. But **Field** has no initial object, since there are no maps between fields of different characteristic. Further examples of nonexistence of adjoints can be found in Exercise 6.3.21.

Adjoint functor theorems

Every functor with a left adjoint preserves limits, but limit-preservation alone does not guarantee the existence of a left adjoint. For example, let $\mathscr{B}$ be any category. The unique functor $\mathscr{B} \to \mathbf{1}$ always preserves limits, but by Example 2.1.9, it only has a left adjoint if $\mathscr{B}$ has an initial object.

On the other hand, if we have a limit-preserving functor $G: \mathscr{B} \to \mathscr{A}$ and $\mathscr{B}$ has all limits, then there is an excellent chance that G has a left adjoint. It is still not always true, but counterexamples are harder to find. For instance (taking $\mathscr{A} = \mathbf{1}$ again), can you find a category $\mathscr{B}$ that has all limits but no initial object?

The condition of having all limits is so important that it has its own word:

Definition 6.3.6 A category is **complete** (or properly, **small complete**) if it has all limits.

There are various results called adjoint functor theorems, all of the following form:

> Let $\mathscr{A}$ be a category, $\mathscr{B}$ a complete category, and $G: \mathscr{B} \to \mathscr{A}$ a functor. Suppose that $\mathscr{A}$, $\mathscr{B}$ and G satisfy certain further conditions. Then
>
> $\qquad$ G has a left adjoint $\iff$ G preserves limits.

The forwards implication is immediate from Theorem 6.3.1. It is the backwards implication that concerns us here.

Typically, the 'further conditions' involve the distinction between small and large collections. But there is a special case in which these complications disappear, and I will use it to explain the main idea behind the proofs of the adjoint functor theorems. It is the case where the categories $\mathscr{A}$ and $\mathscr{B}$ are ordered sets.

As we saw in Section 5.1, limits in ordered sets are meets. More precisely, if $D\colon \mathbf{I} \to \mathbf{B}$ is a diagram in an ordered set $\mathbf{B}$, then

$$\lim_{\leftarrow \mathbf{I}} D = \bigwedge_{I \in \mathbf{I}} D(I),$$

with one side defined if and only if the other is. So an ordered set is complete if and only if every subset has a meet. Similarly, a map $G\colon \mathbf{B} \to \mathbf{A}$ of ordered sets preserves limits if and only if

$$G\left(\bigwedge_{i \in I} B_i\right) = \bigwedge_{i \in I} G(B_i)$$

whenever $(B_i)_{i \in I}$ is a family of elements of $\mathbf{B}$ for which a meet exists.

We now show that for ordered sets, there is an adjoint functor theorem of the simplest possible kind: there are no 'further conditions' at all.

Proposition 6.3.7 (Adjoint functor theorem for ordered sets)   *Let* $\mathbf{A}$ *be an ordered set,* $\mathbf{B}$ *a complete ordered set, and* $G\colon \mathbf{B} \to \mathbf{A}$ *an order-preserving map. Then*

$$G \text{ has a left adjoint} \iff G \text{ preserves meets.}$$

Proof   Suppose that G preserves meets. By Corollary 2.3.7, it is enough to show that for each $A \in \mathbf{A}$, the comma category $(A \Rightarrow G)$ has an initial object. Let $A \in \mathbf{A}$. Then $(A \Rightarrow G)$ is an ordered set, namely, $\{B \in \mathbf{B} \mid A \leq G(B)\}$ with the order inherited from $\mathbf{B}$. We have to show that $(A \Rightarrow G)$ has a least element.

Since $\mathbf{B}$ is complete, the meet $\bigwedge_{B \in \mathbf{B}\colon A \leq G(B)} B$ exists in $\mathbf{B}$. This is the meet of all the elements of $(A \Rightarrow G)$, so it suffices to show that the meet is itself an element of $(A \Rightarrow G)$. And indeed, since G preserves meets, we have

$$G\left(\bigwedge_{B \in \mathbf{B}\colon A \leq G(B)} B\right) = \bigwedge_{B \in \mathbf{B}\colon A \leq G(B)} G(B) \geq A,$$

as required.   □

In the general setting of Corollary 2.3.7, the initial object of $(A \Rightarrow G)$ is the pair $\left(F(A), A \xrightarrow{\eta_A} GF(A)\right)$, where F is the left adjoint and η is the unit map. So in Proposition 6.3.7, the left adjoint F is given by

$$F(A) = \bigwedge_{B \in \mathbf{B}\colon A \leq G(B)} B. \tag{6.19}$$

Example 6.3.8   Consider Proposition 6.3.7 in the case $\mathbf{A} = \mathbf{1}$. The unique functor $G: \mathbf{B} \to \mathbf{1}$ automatically preserves meets, and, as observed above, a left adjoint to G is an initial object of $\mathbf{B}$. So in the case $\mathbf{A} = \mathbf{1}$, the proposition states that a complete ordered set has a least element. This is not quite trivial, since completeness means the existence of all meets, whereas a least element is an empty *join*.

By (6.19), the least element of $\mathbf{B}$ is $\bigwedge_{B \in \mathbf{B}} B$. Thus, a least element is not only a colimit of the functor $\emptyset \to \mathbf{B}$; it is also a limit of the identity functor $\mathbf{B} \to \mathbf{B}$.

The synonym 'least upper bound' for 'join' suggests a theorem: that a poset with all meets also has all joins. Indeed, given a poset $\mathbf{B}$ with all meets, the join of a subset of $\mathbf{B}$ is simply the meet of its upper bounds: quite literally, its least upper bound.

Let us now attempt to extend Proposition 6.3.7 from ordered sets to categories, starting with a limit-preserving functor G from a complete category $\mathscr{B}$ to a category $\mathscr{A}$. In the case of ordered sets, we had for each $A \in \mathscr{A}$ an inclusion map $P_A : (A \Rightarrow G) \hookrightarrow \mathbf{B}$, and we showed that the left adjoint F was given by

$$F(A) = \lim_{\leftarrow (A \Rightarrow G)} P_A. \tag{6.20}$$

In the general case, the analogue of the inclusion functor is the projection functor

$$
\begin{aligned}
P_A : \quad & (A \Rightarrow G) & \to & \quad \mathscr{B} \\
& \left(B, A \xrightarrow{f} G(B) \right) & \mapsto & \quad B.
\end{aligned}
\tag{6.21}
$$

The case of ordered sets suggests that in general, equation (6.20) might define a left adjoint F to G. And indeed, it can be shown that if this limit in $\mathscr{B}$ exists and is preserved by G, then (6.20) really does give a left adjoint (Theorem X.1.2 of Mac Lane (1971)).

This might seem to suggest that our adjoint functor theorem generalizes smoothly from ordered sets to arbitrary categories, with no need for further conditions. But it does not, for reasons that are quite subtle.

Those reasons are more easily explained if we relax our terminology slightly. When we defined limits, we built in the condition that the shape category $\mathbf{I}$ was small. However, the definition of limit makes sense for an arbitrary category $\mathbf{I}$. In this discussion, we will need to refer to this more inclusive notion of limit, so let us temporarily suspend the convention that the shape categories $\mathbf{I}$ of limits are always small.

Now, in the template for adjoint functor theorems stated above (after Definition 6.3.6), it was only required that $\mathscr{B}$ has, and G preserves, *small* limits. But if $\mathscr{B}$ is a large category then $(A \Rightarrow G)$ might also be large, since to specify an object or map in $(A \Rightarrow G)$, we have to specify (among other things) an object or map in $\mathscr{B}$. So, the limit (6.20) defining the left adjoint is not guaranteed to be small. Hence there is no guarantee that this limit exists in $\mathscr{B}$, nor that it is preserved by G. It follows that the functor F 'defined' by (6.20) might not be defined at all, let alone a left adjoint.

(The reader experiencing difficulty with reasoning about small and large collections might usefully compare finite and infinite collections. For instance, if $\mathscr{B}$ is a finite category and $\mathscr{A}$ has finite hom-sets then $(A \Rightarrow G)$ is also finite, but otherwise $(A \Rightarrow G)$ might be infinite.)

Proposition 6.3.7 still stands, since there we were dealing with ordered *sets*, which as categories are small. We might hope to extend it from posets to arbitrary small categories, since the problem just described affects only large categories. But this turns out not to be very fruitful, since in fact, complete posets are the *only* complete small categories (Exercise 6.3.23).

Alternatively, we could try to salvage the argument by assuming that $\mathscr{B}$ has, and G preserves, *all* (possibly large) limits. But again, this is unhelpful: there are almost no such categories $\mathscr{B}$.

The situation therefore becomes more complicated. Each of the best-known adjoint functor theorems imposes further conditions implying that the large limit $\lim_{\leftarrow(A \Rightarrow G)} P_A$ can be replaced by a small limit in some clever way. This allows one to proceed with the argument above.

The two most famous adjoint functor theorems are the 'general' and the 'special'. Their exact statements and proofs are perhaps less significant than their consequences.

Definition 6.3.9 Let $\mathscr{C}$ be a category. A **weakly initial set** in $\mathscr{C}$ is a set $\mathbf{S}$ of objects with the property that for each $C \in \mathscr{C}$, there exist an element $S \in \mathbf{S}$ and a map $S \to C$.

Note that $\mathbf{S}$ must be a set, that is, small. So, the existence of a weakly initial set is some kind of size restriction. Such size restrictions are comparable to finiteness conditions in algebra.

Theorem 6.3.10 (General adjoint functor theorem) *Let $\mathscr{A}$ be a category, $\mathscr{B}$ a complete category, and $G \colon \mathscr{B} \to \mathscr{A}$ a functor. Suppose that $\mathscr{B}$ is locally small and that for each $A \in \mathscr{A}$, the category $(A \Rightarrow G)$ has a weakly initial set. Then*

$$G \text{ has a left adjoint} \iff G \text{ preserves limits.}$$

Proof See the appendix.   □

Example 6.3.11 The general adjoint functor theorem (GAFT) implies that for any category $\mathscr{B}$ of algebras (**Grp**, **Vect**$_k$, ...), the forgetful functor U: $\mathscr{B} \to$ **Set** has a left adjoint. Indeed, we saw in Example 5.1.23 that $\mathscr{B}$ has all limits, and in Example 5.3.4 that U preserves them. Also, $\mathscr{B}$ is locally small. To apply GAFT, we now just have to check that for each $A \in$ **Set**, the comma category $(A \Rightarrow U)$ has a weakly initial set. This requires a little cardinal arithmetic, omitted here; see Exercise 6.3.24.

So GAFT tells us that, for instance, the free group functor exists. In Examples 1.2.4(a) and 2.1.3(b), we began to see the trickiness of explicitly constructing the free group on a generating set A. One has to define the set of 'formal expressions' (such as $x^{-1}yx^2zy^{-3}$, with $x, y, z \in A$), then say what it means for two such expressions to be equivalent (so that $x^{-2}x^5y$ is equivalent to x^3y), then define $F(A)$ to be the set of all equivalence classes, then define the group structure, then check the group axioms, then prove that the resulting group has the universal property required. But using GAFT, we can avoid these complications entirely.

The price to be paid is that GAFT does not give us an explicit description of free groups (or left adjoints more generally). When people speak of knowing some object 'explicitly', they usually mean knowing its elements. An element of an object is a map *into* it, and we have no handle on maps into $F(A)$: since F is a left adjoint, it is maps *out* of $F(A)$ that we know about. This is why explicit descriptions of left adjoints are often hard to come by.

Example 6.3.12 More generally, GAFT guarantees that forgetful functors between categories of algebras, such as

$$\textbf{Ab} \to \textbf{Grp}, \quad \textbf{Grp} \to \textbf{Mon}, \quad \textbf{Ring} \to \textbf{Mon}, \quad \textbf{Vect}_{\mathbb{C}} \to \textbf{Vect}_{\mathbb{R}},$$

have left adjoints. (Some of them are described in Examples 2.1.3.) This is 'more generally' because **Set** can be seen as a degenerate example of a category of algebras, in the sense of Remark 2.1.4: a group, ring, etc., is a set equipped with some operations satisfying some equations, and a set is a set equipped with no operations satisfying no equations.

The special adjoint functor theorem (SAFT) operates under much tighter hypotheses than GAFT, and is much less widely applicable. Its main advantage is that it removes the condition on weakly initial sets. Indeed, it removes *all* further conditions on the functor G.

Theorem 6.3.13 (Special adjoint functor theorem) *Let $\mathscr{A}$ be a category, $\mathscr{B}$ a complete category, and $G: \mathscr{B} \to \mathscr{A}$ a functor. Suppose that $\mathscr{A}$ and $\mathscr{B}$*

are locally small, and that $\mathscr{B}$ satisfies certain further conditions. Then

$$G \text{ has a left adjoint} \iff G \text{ preserves limits.}$$

A precise statement and proof can be found in Section V.8 of Mac Lane (1971).

Example 6.3.14   Here is the classic application of SAFT. Let **CptHff** be the category of compact Hausdorff spaces, and U: **CptHff** $\to$ **Top** the forgetful functor. SAFT tells us that U has a left adjoint F, turning any space into a compact Hausdorff space in a canonical way.

The existence of this left adjoint is far from obvious, and verifying the hypotheses of SAFT (or indeed, constructing F in any other way) requires some deep theorems of topology. Given a space X, the resulting compact Hausdorff space $F(X)$ is called its **Stone–Čech compactification**. Provided that X satisfies some mild separation conditions, the unit of the adjunction at X is an embedding, so that $UF(X)$ contains X as a subspace.

Another advantage of SAFT is that one can extract from its proof a fairly explicit formula for the left adjoint. In this case, it tells us that $F(X)$ is the closure of the image of the canonical map

$$X \to [0, 1]^{\mathbf{Top}(X, [0,1])},$$

where the codomain is a power of $[0, 1]$ in **Top**.

Cartesian closed categories

We have seen that for every set B, there is an adjunction $(- \times B) \dashv (-)^B$ (Example 2.1.6), and that for every category $\mathscr{B}$, there is an adjunction $(- \times \mathscr{B}) \dashv [\mathscr{B}, -]$ (Remark 4.1.23(c)).

Definition 6.3.15   A category $\mathscr{A}$ is **cartesian closed** if it has finite products and for each $B \in \mathscr{A}$, the functor $- \times B \colon \mathscr{A} \to \mathscr{A}$ has a right adjoint.

We write the right adjoint as $(-)^B$, and, for $C \in \mathscr{A}$, call C^B an **exponential**. We may think of C^B as the space of maps from B to C. Adjointness says that for all $A, B, C \in \mathscr{A}$,

$$\mathscr{A}(A \times B, C) \cong \mathscr{A}(A, C^B)$$

naturally in A and C. In fact, the isomorphism is natural in B too; that comes for free.

Example 6.3.16   **Set** is cartesian closed; C^B is the function set **Set**(B, C).

Example 6.3.17   **CAT** is cartesian closed; $\mathscr{C}^{\mathscr{B}}$ is the functor category $[\mathscr{B}, \mathscr{C}]$.

In any cartesian closed category with finite sums, the isomorphisms (6.17) and (6.18) of Example 6.3.3 hold, for the same reasons as stated there. The objects of a cartesian closed category therefore possess an arithmetic like that of the natural numbers. This thought can be developed in several interesting directions, but here we just note that these isomorphisms provide a way of proving that a category is *not* cartesian closed.

Example 6.3.18 $\mathbf{Vect}_k$ is not cartesian closed, for any field k. It does have finite products, as we saw in Example 5.1.5: binary product is direct sum $\oplus$, and the terminal object is the trivial vector space $\{0\}$, which is also initial. But if $\mathbf{Vect}_k$ were cartesian closed then equations (6.17) would hold, so that $\{0\} \oplus B \cong \{0\}$ for all vector spaces B. This is plainly false.

Remark 6.3.19 For any vector spaces V and W, the set $\mathbf{Vect}_k(V, W)$ of linear maps can itself be given the structure of a vector space, as in Example 1.2.12. Let us now call this vector space $[V, W]$.

Given that exponentials are supposed to be 'spaces of maps', you might expect $\mathbf{Vect}_k$ to be cartesian closed, with $[-, -]$ as its exponential. We have just seen that this cannot be so. But as it turns out, the linear maps $U \to [V, W]$ correspond to the *bi*linear maps $U \times V \to W$, or equivalently the linear maps $U \otimes V \to W$. In the jargon, $\mathbf{Vect}_k$ is an example of a 'monoidal closed category'. These are like cartesian closed categories, but with the cartesian (categorical) product replaced by some other operation called 'product', in this case the tensor product of vector spaces.

For any set I, the product category $\mathbf{Set}^I$ is cartesian closed, just because $\mathbf{Set}$ is. (Exponentials in $\mathbf{Set}^I$, as well as products, are computed pointwise.) Put another way, $[\mathbf{A}^{\mathrm{op}}, \mathbf{Set}]$ is cartesian closed whenever $\mathbf{A}$ is discrete. We now show that, in fact, $[\mathbf{A}^{\mathrm{op}}, \mathbf{Set}]$ is cartesian closed for any small category $\mathbf{A}$ whatsoever.

In preparation for proving this, let us conduct a thought experiment. Write $\hat{\mathbf{A}} = [\mathbf{A}^{\mathrm{op}}, \mathbf{Set}]$. If $\hat{\mathbf{A}}$ *is* cartesian closed, what must exponentials in $\hat{\mathbf{A}}$ be? In other words, given presheaves Y and Z, what must Z^Y be in order that

$$\hat{\mathbf{A}}(X, Z^Y) \cong \hat{\mathbf{A}}(X \times Y, Z) \tag{6.22}$$

for all presheaves X? If this is true for all presheaves X, then in particular it is true when X is representable, so

$$Z^Y(A) \cong \hat{\mathbf{A}}(H_A, Z^Y) \cong \hat{\mathbf{A}}(H_A \times Y, Z)$$

for all $A \in \mathbf{A}$, the first step by Yoneda. This tells us what Z^Y must be. Notice that $Z^Y(A)$ is not simply $Z(A)^{Y(A)}$, as one might at first guess: exponentials in a presheaf category are *not* generally computed pointwise.

Theorem 6.3.20   *For any small category* **A**, *the presheaf category* $\hat{\mathbf{A}}$ *is cartesian closed.*

Here is the strategy of the proof. The argument in the thought experiment gives us the isomorphism (6.22) whenever X is representable. A general presheaf X is not representable, but it is a colimit of representables, and this allows us to bootstrap our way up.

Proof   We know that $\hat{\mathbf{A}}$ has all limits, and in particular, finite products. It remains to show that $\hat{\mathbf{A}}$ has exponentials. Fix $Y \in \hat{\mathbf{A}}$.

First we prove that $- \times Y \colon \hat{\mathbf{A}} \to \hat{\mathbf{A}}$ preserves colimits. (Eventually we will prove that $- \times Y$ has a right adjoint, from which preservation of colimits follows, but our proof that it has a right adjoint will *use* preservation of colimits.) Indeed, since products and colimits in $\hat{\mathbf{A}}$ are computed pointwise, it is enough to prove that for any set S, the functor $- \times S \colon \mathbf{Set} \to \mathbf{Set}$ preserves colimits, and this follows from the fact that **Set** is cartesian closed.

For each presheaf Z on **A**, let Z^Y be the presheaf defined by

$$Z^Y(A) = \hat{\mathbf{A}}(H_A \times Y, Z)$$

for all $A \in \mathbf{A}$. This defines a functor $(-)^Y \colon \hat{\mathbf{A}} \to \hat{\mathbf{A}}$.

I claim that $(- \times Y) \dashv (-)^Y$. Let $X, Z \in \hat{\mathbf{A}}$. Write $P \colon \mathbf{E}(X) \to \mathbf{A}$ for the projection (as in Definition 6.2.16), and write $H_P = H_\bullet \circ P$. Then

$$\hat{\mathbf{A}}(X, Z^Y) \cong \hat{\mathbf{A}}\left(\varinjlim_{\to \mathbf{E}(X)} H_P, Z^Y \right) \tag{6.23}$$

$$\cong \varprojlim_{\leftarrow \mathbf{E}(X)} \hat{\mathbf{A}}(H_P, Z^Y) \tag{6.24}$$

$$\cong \varprojlim_{\leftarrow \mathbf{E}(X)} Z^Y(P) \tag{6.25}$$

$$\cong \varprojlim_{\leftarrow \mathbf{E}(X)} \hat{\mathbf{A}}(H_P \times Y, Z) \tag{6.26}$$

$$\cong \hat{\mathbf{A}}\left(\varinjlim_{\to \mathbf{E}(X)} (H_P \times Y), Z \right) \tag{6.27}$$

$$\cong \hat{\mathbf{A}}\left(\left(\varinjlim_{\to \mathbf{E}(X)} H_P \right) \times Y, Z \right) \tag{6.28}$$

$$\cong \hat{\mathbf{A}}(X \times Y, Z) \tag{6.29}$$

naturally in X and Z. Here (6.23) and (6.29) follow from Theorem 6.2.17; (6.24) and (6.27) are because representables preserve limits (as rephrased in Remark 6.2.3); (6.25) is by Yoneda; (6.26) is by definition of Z^Y; and (6.28) is because $- \times Y$ preserves colimits.   □

This result can be seen as a step along the road to topos theory. A topos is a category with certain special properties. Topos theory unifies, in an extraordinary way, important aspects of logic and geometry.

For instance, a topos can be regarded as a 'universe of sets': **Set** is the most basic example of a topos, and every topos shares enough features with **Set** that one can reason with its objects as if they were sets of some exotic kind. On the other hand, a topos can be regarded as a generalized topological space: every space gives rise to a topos (namely, the category of sheaves on it), and topological properties of the space can be reinterpreted in a useful way as categorical properties of its associated topos.

By definition, a topos is a cartesian closed category with finite limits and with one further property: the existence of a so-called subobject classifier. For example, the two-element set 2 is the subobject classifier of **Set**, which means, informally, that subsets of a set A correspond one-to-one with maps $A \to 2$. Exercises 6.3.26 and 6.3.27 give the formal definition of subobject classifier, then guide you through the proof that **Set**, and, more generally, every presheaf category, is a topos.

Exercises

6.3.21 (a) Prove that the forgetful functor $U \colon \textbf{Grp} \to \textbf{Set}$ has no right adjoint.

(b) Prove that the chain of adjunctions $C \dashv D \dashv O \dashv I$ in Exercise 3.2.16 extends no further in either direction.

(c) Does the chain of adjunctions in Exercise 2.1.17 extend further in either direction?

6.3.22 Let $\mathscr{A}$ be a locally small category. For functors $U \colon \mathscr{A} \to \textbf{Set}$, consider the following three conditions: (A) U has a left adjoint; (R) U is representable; (L) U preserves limits.

(a) Show that (A) $\Longrightarrow$ (R) $\Longrightarrow$ (L).

(b) Show that if $\mathscr{A}$ has sums then (R) $\Longrightarrow$ (A).

(If $\mathscr{A}$ satisfies the hypotheses of the special adjoint functor theorem then also (L) $\Longrightarrow$ (A), so the three conditions are equivalent.)

6.3.23 (a) Prove that every preordered set is equivalent (as a category) to an ordered set.

(b) Let $\mathscr{A}$ be a category with all small products. Suppose that $\mathscr{A}$ is not a preorder, so that there exists a parallel pair of maps $A \underset{g}{\overset{f}{\rightrightarrows}} B$ in $\mathscr{A}$ with

$f \neq g$. By considering the maps $A \to B^I$ for each set I, prove that $\mathscr{A}$ is not small.

(c) Deduce that every small category with small products is equivalent to a complete ordered set.

(d) Adapt the argument to prove that every finite category with finite products is equivalent to a complete ordered set.

6.3.24 Probably the most important application of the general adjoint functor theorem is to proving that forgetful functors between categories of algebras have left adjoints (Example 6.3.11). Verifying the hypotheses requires some cardinal arithmetic. Here is a typical example.

(a) Let A be a set. Prove that for any group G and family $(g_a)_{a \in A}$ of elements of G, the subgroup of G generated by $\{g_a \mid a \in A\}$ has cardinality at most $\max\{|\mathbb{N}|, |A|\}$.

(b) Prove that for any set S, the collection of isomorphism classes of groups of cardinality at most $|S|$ is small.

(c) Let $U \colon \mathbf{Grp} \to \mathbf{Set}$ be the forgetful functor from groups to sets. Deduce from (a) and (b) that for every set A, the comma category $(A \Rightarrow U)$ has a weakly initial set.

(d) Use GAFT to conclude that U has a left adjoint.

6.3.25 Let $\mathbf{A}$ be a small cartesian closed category. Prove that the Yoneda embedding $\mathbf{A} \to [\mathbf{A}^{\mathrm{op}}, \mathbf{Set}]$ preserves the whole cartesian closed structure (exponentials as well as products).

6.3.26 Recall from Exercise 5.1.40 the notion of subobject. A category $\mathscr{A}$ is **well-powered** if for each $A \in \mathscr{A}$, the class of subobjects of A is small, that is, a set. (All of our usual examples of categories are well-powered.) Let $\mathscr{A}$ be a well-powered category with pullbacks, and write $\mathrm{Sub}(A)$ for the set of subobjects of an object $A \in \mathscr{A}$.

(a) Deduce from Exercise 5.1.42 that any map $A' \xrightarrow{f} A$ in $\mathscr{A}$ induces a map $\mathrm{Sub}(f) \colon \mathrm{Sub}(A) \to \mathrm{Sub}(A')$.

(b) Show that this determines a functor $\mathrm{Sub} \colon \mathscr{A}^{\mathrm{op}} \to \mathbf{Set}$. (Hint: use Exercise 5.1.35.)

(c) For some categories $\mathscr{A}$, the functor Sub is representable. A **subobject classifier** for $\mathscr{A}$ is an object $\Omega \in \mathscr{A}$ such that $\mathrm{Sub} \cong H_\Omega$. Prove that 2 is a subobject classifier for $\mathbf{Set}$.

A **topos** is a cartesian closed category with finite limits and a subobject classifier. You have just completed the proof that $\mathbf{Set}$ is a topos.

6.3.27 This exercise follows on from the last, culminating in the proof that every presheaf category is a topos. Let **A** be a small category.

(a) By conducting a thought experiment similar to the one before the statement of Theorem 6.3.20, find out what the subobject classifier Ω of $[\mathbf{A}^{op}, \mathbf{Set}]$ must be if it exists.

(b) Prove that this Ω is indeed a subobject classifier.

(c) Conclude that $[\mathbf{A}^{op}, \mathbf{Set}]$ is a topos.

Appendix

Proof of the general adjoint functor theorem

Here we prove the general adjoint functor theorem, which for convenience is restated below. The left-to-right implication follows immediately from Theorem 6.3.1; it is the right-to-left implication that we have to prove.

Theorem 6.3.10 (General adjoint functor theorem) *Let $\mathscr{A}$ be a category, $\mathscr{B}$ a complete category, and $G\colon \mathscr{B} \to \mathscr{A}$ a functor. Suppose that $\mathscr{B}$ is locally small and that for each $A \in \mathscr{A}$, the category $(A \Rightarrow G)$ has a weakly initial set. Then*

$$G \text{ has a left adjoint} \iff G \text{ preserves limits.}$$

The heart of the proof is the case $\mathscr{A} = \mathbf{1}$, where GAFT asserts that a complete locally small category with a weakly initial set has an initial object. We prove this first.

The proof of this special case is illuminated by considering the even more special case where $\mathscr{A} = \mathbf{1}$ and the category $\mathscr{B}$ is a poset $\mathbf{B}$. We saw in Example 6.3.8 that the initial object (least element) of a complete poset $\mathbf{B}$ can be constructed as the meet of all its elements. Otherwise put, it is the limit of the identity functor $1_\mathbf{B}\colon \mathbf{B} \to \mathbf{B}$.

One might try to extend this result to arbitrary categories $\mathscr{B}$ by proving that the limit of the identity functor $1_\mathscr{B}\colon \mathscr{B} \to \mathscr{B}$ is (if it exists) an initial object. This is indeed true (Exercise A.3 below). However, it is unhelpful: for if $\mathscr{B}$ is large then the limit of $1_\mathscr{B}$ is a large limit, but we are only given that $\mathscr{B}$ has small limits.

We seem to be at an impasse – but this is where the clever idea behind GAFT comes in. In order to construct the least element of a complete poset, it is not necessary to take the meet of *all* the elements. More economically, we could just take the meet of the elements of some weakly initial subset (Exercise A.4).

In general, for an arbitrary complete category, the limit of any weakly initial set is an initial object. We prove this now.

Lemma A.1　*Let $\mathscr{C}$ be a complete locally small category with a weakly initial set. Then $\mathscr{C}$ has an initial object.*

Proof　Let **S** be a weakly initial set in $\mathscr{C}$. Regard **S** as a full subcategory of $\mathscr{C}$; then **S** is small, since $\mathscr{C}$ is locally small. We may therefore take a limit cone

$$\left(0 \xrightarrow{\ p_S\ } S\right)_{S \in \mathbf{S}} \tag{A.1}$$

of the inclusion $\mathbf{S} \hookrightarrow \mathscr{C}$. We prove that 0 is initial.

Let $C \in \mathscr{C}$. We have to show that there is exactly one map $0 \to C$. Certainly there is at least one, since we may choose some $S \in \mathbf{S}$ and map $j: S \to C$, and we then have the composite $jp_S: 0 \to C$. To prove uniqueness, let $f, g: 0 \to C$. Form the equalizer

$$E \xrightarrow{\ i\ } 0 \underset{g}{\overset{f}{\rightrightarrows}} C.$$

Since **S** is weakly initial, we may choose $S \in \mathbf{S}$ and $h: S \to E$. We then have maps

$$0 \xrightarrow{\ p_S\ } S \xrightarrow{\ h\ } E \xrightarrow{\ i\ } 0$$

with the property that for all $S' \in \mathbf{S}$,

$$p_{S'}(ihp_S) = (p_{S'}ih)p_S = p_{S'} = p_{S'}1_0$$

(where the second equality follows from (A.1) being a cone). But (A.1) is a *limit* cone, so $ihp_S = 1_0$ by Exercise 5.1.36(a). Hence

$$f = fihp_S = gihp_S = g,$$

as required.　　　　　　　　　　　　　　　　　　　　　　　　　　　□

We have now proved GAFT in the special case $\mathscr{A} = \mathbf{1}$. The rest of the proof is comparatively routine.

Lemma A.2　*Let $\mathscr{A}$ and $\mathscr{B}$ be categories. Let $G: \mathscr{B} \to \mathscr{A}$ be a functor that preserves limits. Then the projection functor $P_A: (A \Rightarrow G) \to \mathscr{B}$ of (6.21) creates limits, for each $A \in \mathscr{A}$. In particular, if $\mathscr{B}$ is complete then so is each comma category $(A \Rightarrow G)$.*

Proof　The first statement is Exercise A.5(b), and the second follows from Lemma 5.3.6.　　　　　　　　　　　　　　　　　　　　　□

We now prove GAFT. By Corollary 2.3.7, it is enough to show that $(A \Rightarrow G)$ has an initial object for each $A \in \mathscr{A}$. Let $A \in \mathscr{A}$. By Lemma A.2, $(A \Rightarrow G)$ is complete, and by hypothesis, it has a weakly initial set. It is also locally small, since $\mathscr{B}$ is. Hence by Lemma A.1, it has an initial object, as required.

Exercises

A.3   In this exercise, we suspend the convention (made implicitly in Definition 5.1.19) that we only speak of the limit of a functor $\mathbf{I} \to \mathscr{C}$ when $\mathbf{I}$ is small. Let $\mathscr{B}$ be a category, possibly large. The aim is to prove that a limit of the identity functor on $\mathscr{B}$ is exactly an initial object of $\mathscr{B}$.

(a) Let 0 be an initial object of $\mathscr{B}$. Show that the cone $(0 \to B)_{B \in \mathscr{B}}$ on the identity functor $1_{\mathscr{B}}$ is a limit cone.

(b) Now let $\left(L \xrightarrow{p_B} B \right)_{B \in \mathscr{B}}$ be a limit cone on $1_{\mathscr{B}}$. Prove that p_L is the identity on L, and deduce that L is initial.

A.4   Here you will prove the special case of Lemma A.1 in which the category concerned is a poset. Let C be a poset and $S \subseteq C$.

(a) What does it mean, in purely order-theoretic terms, for S to be a weakly initial set in C?

(b) Prove directly that if S is weakly initial and the meet $\bigwedge_{s \in S} s$ exists then $\bigwedge_{s \in S} s$ is a least element of C.

A.5   Let $G \colon \mathscr{B} \to \mathscr{A}$ be a limit-preserving functor, and let $A \in \mathscr{A}$.

(a) Show that for any small category $\mathbf{I}$, a diagram of shape $\mathbf{I}$ in $(A \Rightarrow G)$ amounts to a diagram E of shape $\mathbf{I}$ in $\mathscr{B}$ together with a cone on $G \circ E$ with vertex A.

(b) Prove that the projection functor $P_A \colon (A \Rightarrow G) \to \mathscr{B}$ of (6.21) creates limits.

Further reading

This book is intentionally short. Even some topics that are included in most introductions to category theory are omitted here. I will indicate some of the topics that lie beyond the scope of this book, and suggest where you might read about them. Since there is far more written on category theory than anyone could read in a lifetime, these recommendations are necessarily subjective.

The towering presence among category theory books is the classic by one of its founders:

> Saunders Mac Lane, *Categories for the Working Mathematician.*
> Springer, 1971; second edition with two new chapters, 1998.

It is so well-written that more than forty years on, it is still the most popular introduction to the subject. It addresses a more mature readership than this text, and covers many topics omitted here, including monads (one formalization of the idea of algebraic theory), monoidal categories (categories equipped with a tensor product), 2-categories (mentioned at the end of our Chapter 1), abelian categories (categories of modules), ends (an elegant generalization of the notion of limit), and Kan extensions (which provide the tongue-in-cheek title of the book's final section: 'All concepts are Kan extensions').

Another well-liked book, longer than the one you hold in your hands but written for a similar readership, is:

> Steve Awodey, *Category Theory*. Oxford University Press, 2010.

Awodey's book covers less than Mac Lane's, but is particularly strong on connections between category theory and other parts of logic. It has a full chapter on cartesian closed categories, and also covers the theory of monads.

Those who prefer lectures to books might try this library of 75 ten-minute introductory category theory videos:

Eugenia Cheng and Simon Willerton, The Catsters. Available at www.youtube.com/user/TheCatsters, 2007–2010.

Other than the topics treated here, they cover monads, enriched categories, internal groups (and other internal algebraic structures), string diagrams (which we touched on in Remark 2.2.9), and several more sophisticated topics.

For inspiration as much as instruction, here are two further recommendations.

Saunders Mac Lane, *Mathematics: Form and Function*. Springer, 1986.

F. William Lawvere and Stephen H. Schanuel, *Conceptual Mathematics: A First Introduction to Categories*. Cambridge University Press, 1997.

Mathematics: Form and Function is a tour through much of pure and applied mathematics, written from a categorical perspective. Its declared purpose is to present the author's philosophy of mathematics, but it can also be enjoyed for its many excellent vignettes of exposition. (Beware of the numerous small errors.) *Conceptual Mathematics* is a thought-provoking text and an intriguing experiment: category theory for high-school students, complete with classroom dialogues.

For categorical topics beyond the scope of this book, two good general references are:

Francis Borceux, *Handbook of Categorical Algebra, Volumes 1–3*. Cambridge University Press, 1994.

Various authors, *The nLab*. Available at http://ncatlab.org, 2008– present.

Borceux's encyclopaedic work often takes a different point of view from the present text, but covers many, many more topics. Apart from those just mentioned in connection with other books, some of the more important ones are fibrations, bimodules (also called profunctors or distributors), Lawvere theories, Cauchy completeness, Morita equivalence, absolute colimits, and flatness.

The *n*Lab is an ever-growing online resource for mathematics, focusing on category theory and operating on similar principles to Wikipedia. Individual entries can be idiosyncratic, but it has become a very useful reference for advanced categorical topics.

Vigorous research in category theory continues to be done. The sources listed above provide ample onward references for anyone wishing to explore.

Other texts cited

Timothy Gowers, *Mathematics: A Very Short Introduction.* Oxford University Press, 2002.

G. M. Kelly, *Basic Concepts of Enriched Category Theory.* Cambridge University Press, 1982. Also *Reprints in Theory and Applications of Categories* 10 (2005), 1–136, available at www.tac.mta.ca/tac/reprints.

F. William Lawvere and Robert Rosebrugh, *Sets for Mathematics.* Cambridge University Press, 2003.

Tom Leinster, Rethinking set theory. *American Mathematical Monthly*, to appear (2014). Also available at http://arxiv.org/abs/1212.6543.

Index of notation

Index

Printed in the United States
By Bookmasters